Foundations of Ultraprecision Mecha

Developments in Nanotechnology

Series Editor: D. Keith Bowen, University of Warwick, UK

This book is part of a series. The publisher will accept continuation orders which may be cancelled at any time and which provide for automatic billing and shipping of each title in the series upon publication. Please write for details.

Foundations of Ultraprecision Mechanism Design

S.T. Smith
University of North Carolina at Charlotte, USA

and

D.G. Chetwynd
University of Warwick, UK

GORDON AND BREACH SCIENCE PUBLISHERS
Australia • Canada • China • France • Germany • India • Japan • Luxembourg
Malaysia • The Netherlands • Russia • Singapore • Switzerland

First published 1992
Second printing with corrections 1994
Third printing 1997
Fourth printing 1998

Amsteldijk 166
1st Floor
1079 LH Amsterdam
The Netherlands

Transferred to Digital Printing 2003

Cover design © 1992 by Julie Hathaway

British Library/Library of Congress Cataloging-in-Publication Data

Smith, S.T. (Stuart T.), 1961–
      Foundations of ultraprecision mechanism design / S.T. Smith and D.G. Chetwynd
         p. cm.– (Developments in nanotechnology; v. 2)
      Includes bibliographical references and index.
      ISBN 2-88124-840-3 (hard) 2-88449-001-9 (soft)
      1. Machine design. 2. Mechanical movements. I. Chetwynd, D.G. (Derek G.), 198- . II. Title. III. Series.
TJ233.S65 1992
621.8′15--dc20   92–18346
                                                              CIP

ISSN: 1053-7465
ISBN: 2-88449-001-9 (softcover)

CONTENTS

CHAPTER 5
DRIVE COUPLINGS AND THE MECHANICS OF CONTACT — 131

CHAPTER 6
LEVER MECHANISMS OF HIGH RESOLUTION — 155

CHAPTER 9
SLIDEWAYS FOR LONG RANGE PRECISION MOTION   255

INTRODUCTION
TO THE SERIES

The nanotechnological revolution will be as pervasive as the first industrial revolution, for it also introduces concepts and techniques that will influence virtually all aspects of technology and manufacture. Nanotechnology is concerned with the design and manufacture of components with submicrometre dimensions, of larger components with submicrometre tolerances or surface finishes, and of machines with submicrometre precision of positioning or motion. The achievement of ultra-high precision in machining and the manufacture of extremely small devices opens up prospects as diverse and futuristic as massive computing power, medical diagnostic and therapeutic devices inside the bloodstream, global personal communications, still smaller and higher resolution optical devices and high economy, low cost automobiles. The economy of automobile engines has been steadily improving as the precision of component manufacture has increased. The high quality of videocassette recorders is ensured by the reliable, low-cost production of nanometric surface finishes on the recording heads and precise drive spindles. We are seeing in mechanical engineering the same movement towards miniaturization that has occurred over the last three decades in electronic engineering, and we should learn from the analogy; the nanotechnologist must draw from many disciplines to design with confidence in this field. The compartmentalization and consequent simplification of knowledge that works adequately on the large scale make no sense when applied to, for example, submicrometre 'swarf' from an ultra-high precision machining process.

It takes a long time to establish an enabling technology. It is necessary that the active workers in the field take time to communicate the technology and the excitement they feel in being a part of this revolution, whether they address established engineers or recent graduates. In this series of monographs and books on nanotechnology, we endeavour to stimulate, educate and motivate engineers and scientists at all levels in this new field. It will be cross-disciplinary, dealing with instrumentation and measurement, large-scale manufacture, novel manufacturing methods, applications in medicine, and many other relevant topics. The series is designed to encourage fruitful interactions between those who come from different traditional disciplines.

Whichever discipline it is applied to, nanotechnology depends critically, in devices or development, on highly stable, highly repeatable mechanical systems. The arts of precision mechanics enable the enabling technology and, for this reason, the second book in the series is more concerned with interdisciplinary ideas than it might appear at first glance. Drawing on their many years of experience, the authors have concentrated on the fundamental design principles that good mechanical engineers employ almost sub-consciously and that ought to be more widely appreciated. While, for example, biologists should rightly expect to pass detailed design, stress calculations and such matters to a specialist, it is helpful if they take account of basic design constraints when developing specifications. The book is arranged so that it is useful at this level, while offering deeper insights to those whose task might be to produce movements of a precision best expressed in atoms.

D. Keith Bowen

PREFACE

Is there a special discipline of 'precision mechanical design'? Perhaps not, for good design practice remains good on all scales. On the other hand, the very needs for high precision systems require a certain attitude to design, closely following basic geometrical ideas with fewer than usual compromises for reasons of cost, load carrying capacity or poor environment. There is now wide acceptance that Nanotechnology is a discipline, and that its pursuit makes great demands on mechanical precision. There is also a long tradition of scientific instrument making that has consistently pursued new levels of attainment. Thus we address here, as 'ultraprecision mechanism design', methods essential to modern instruments and nanotechnological machines, but relevant also to less stringent applications. Typically we are concerned with the control of motion or dimension to sub-micrometre or nanometre accuracy, on components of up to hundreds of millimetres in overall size. This is no longer the domain only of research specialists. Compact disc players and video tape recorders require controlled motions of precisions rarely attained twenty years ago. Some of their component tolerances could be achieved only with special skills and much effort. Yet today they are mass produced cheaply enough to incorporate into consumer goods. Routine production of integrated circuits takes for granted that ten or more masks can be aligned sequentially above a wafer with such precision that sub-micrometre features are clearly defined. The next generation of systems may well involve position control to sub-atomic resolution!

The lack of a widely accepted definition for our topic is a bonus because we may then choose our own interpretation and so write of things that we find interesting and challenging. It is also a burden because readers might be disappointed that we have excluded ideas relevant to their interpretation. Our solution to this dilemma is to dwell mainly on basic principles that will apply widely, while including a selection of novel and unusual ideas. We also concentrate, although not exclusively, on design approaches that can be implemented by conventional machining. We readily acknowledge the growing importance of electronics, computing and micromachining, for example, but feel that this is not the place to treat them in detail. We have been satisfied to attempt to bring together the main tricks of the trade of the instrument engineer.

The book is aimed primarily at practicing engineers, who should cope comfortably with all the principles used. Since, particularly, instrument designers may have diverse original backgrounds, it includes a brief review of the analytical techniques and mathematical methods that are used regularly. More experienced readers are asked to forgive such inclusions. We have found that mathematical descriptions sometimes distract attention from the underlying physical concepts upon an understanding of which design depends. For this reason mathematical treatments tend to be concentrated in certain parts of the text. Case studies are used to clarify how the ideas are applied in practice. We wish to encourage invention, the seeking of the ever elusive goal of perfection through a creative application of sound principles. It is a not a conventional textbook on design. The book is intended for practical use and is written to allow readers to dip in wherever seems appropriate, with no need to follow a strict order. Nevertheless, its scope is suitable for use in a final year undergraduate or graduate course on the elements of precision design or scientific instrument design.

Following an introduction to the scale of ultraprecision, the rest of Chapter 1 and Chapter 2 introduce some basic analytic tools. Chapter 3 explores basic concepts of precision design. Many are ideas, such as the exploitation of symmetry, that seem little more than common sense and yet are absent from many machines seen around us. Chapter 4 examines in some depth the use of flexure mechanisms to provide precise motion over a restricted range. Chapter 5 starts with a brief review of contact mechanics in order better to appreciate both the limitations of location devices and the errors found in many drive couplings, which constitute its main topic. Generalizations of the lever principle are studied in Chapter 6, before moving on, in Chapter 7, to an examination of actuators and sensors of sub-nanometre capability. In Chapter 8, we reiterate and extend some of the earlier design points in the context of selecting materials suitable for ultra-precise mechanisms. Only then does Chapter 9 consider the provision of way systems for longer range movements. The final chapter gives a moderately detailed discussion of the dynamics of vibration, especially the resonances and isolation of sensitive displacement measuring systems. We believe this ordering of topics to be logical, but other approaches are equally valid. It should prove quite easy for teachers using it as a text to present topics in their preferred order even where different to ours.

The ideas presented in this book have evolved through our experiences and from discussions over many years with many colleagues in many countries. We are most grateful to them all and it would be invidious to name but a few. In covering such a broad range of topics we have reported on the work of many others and we humbly apologize for any errors and omissions that are wholly our own. Comments from readers are welcomed. Our special thanks go to Julie Hathaway, who designed both document and cover as well as producing the final layouts, to Dr Salam Harb for his helpful comments on early drafts and also to our families and friends who have shared the burden of writing this work without experiencing its rewards.

The book was produced completely using personal computer based software. Page design and final layout were implemented using Ventura Publisher (Xerox Corp.) after initial text processing with Microsoft Word. Diagrams were drawn using Autocad and Coreldraw, while graphs were produced through Matlab. Main text and mathematics are in Palatino fonts with Helvetica used for section headings.

Through frequent usage some materials have become associated with a trade name. Where this has occured the designations have been printed in initial capitals.

We have taken the opportunity afforded by reprinting to correct basic errors that appear in the original version. Mainly these are typographical, which is not to say trivial especially if they occur within an equation. A few other ambiguities have also been removed. The constraints of this revision still leave some areas less well explored than we would ideally wish. Our thanks go to everyone who has commented to us about such matters.

Warwick,   **Stuart T. Smith,**
April 1994   **Derek G. Chetwynd**

1 INTRODUCTION

The concept of ultraprecision engineering is illustrated through the scale of some typical processes that might be considered in this context. Technological examples are compared with similarly scaled biological systems in order to demonstrate both the fundamental nature of engineering at this scale and its relationship with nanotechnology. This chapter also provides a convenient place to clarify terminology that will be used throughout this book, by means of a brief review of common instrumentation terms and the nature of accuracy and errors.

1.1 The scale of ultraprecision

An engineering perspective on precision manufacture necessarily looks to a continuing refinement of machining and processing techniques for the overall improvement of products. The initial drive for high precision is firmly rooted in the industrial revolution and exemplified in such names as Henry Maudslay, Joseph Bramah and Joseph Whitworth in the UK and Eli Whitney, Simeon North and Samuel Colt in the USA. In the intervening century and a half the precision of conventional machine tools has improved by orders of magnitude. Today, some types of components can be produced on a routine basis with a precision (ratio of dimension to tolerance) of one part in 100,000 or more. This is an improvement of perhaps as much as one thousand from those early beginnings. Of course, Maudslay and Whitworth were building measuring machines of much better precision, but their methods were hardly routine. Also their work demonstrates the case that improved metrology instruments precede similar improvements in machine tools. Hume, 1980, provides a good review of the early development of modern precision engineering from this perspective.

By comparison with the rates of some other technological advances this is meagre fare. Advances have been consistently impeded by important barriers to ever increasing precision in mechanical manufacture. Working at high levels of precision requires the greatest care in mechanical design perhaps including the use of concepts not hitherto conceived of in normal engineering practice.

The implementation of a novel design idea might wait upon the development of a material having sufficiently well controlled properties. As tolerances or artefact dimensions reduce further, it ceases to be safe to assume that bulk properties remain true and a completely new range of challenges appear. In order to emphasise the present range of demands, consider the logarithmic scale bar shown in Figure 1.1. Below the scale bar are examples drawn from biology. The fly is a familiar and tangible object and the human egg, the largest human cell at a fraction of a millimetre, is still of sufficient size to be visible to the naked eye. When we move to the size of typical cells, of the order of thousandths of a millimetre (that is a few micrometres*) it becomes increasingly difficult to discriminate objects either visually or by touch. Around this point we reach the limit of accuracy in most engineering workshop practices. At the micrometre level we are totally reliant on the use of measuring instruments to detect the presence of material or motions. Consequently the boundary between conventional and precision engineering is often considered to occur where tolerances to better than a few micrometres are specified.

Below one micrometre we start to discover major limitations to the traditional methods of measurement. Hand tools such as micrometers become more difficult to produce with such precision and, anyway, the human hand becomes rather clumsy. Laboratory microscopes are limited in lateral resolution by the wavelength of visible light. Materials related effects such as thermal expansion and deformation under small loads can totally invalidate measurements. Yet it is at the 0.1 μm level that microelectronic fabrication processes must perform on a routine basis. Reducing further to tens of nanometres, the scale of viruses and larger molecular structures, we move towards the limits of performance of some of the most powerful tools for measurement and manipulation. Scanning electron microscopes (SEMs), optical polishing machines and diffraction grating ruling engines all tend to become ineffectual at around 10 nm resolution. Nevertheless, there are numerous examples of everyday, mass produced engineering components which require tolerances at this level. Examples include gauge blocks, diesel fuel injectors and similar hydraulic components, spindles for high speed bearings and video recording heads, magnetic reading heads, force probes, optical lenses and moulds for plastic lenses.

---

* The correct S.I. usage refers to one millionth part of a metre as one micrometre, given the symbol μm. In older texts, and verbally, it is commonly referred to as a micron, sometimes using the symbol μ. Note that the USA spellings use 'meter', 'micrometer', etc. In some parts of the USA and some disciplines the word micron is used to indicate a millionth of an inch. Similarly, in many early British texts the nanometre (10^{-9} m) is referred to by the symbol μμ.

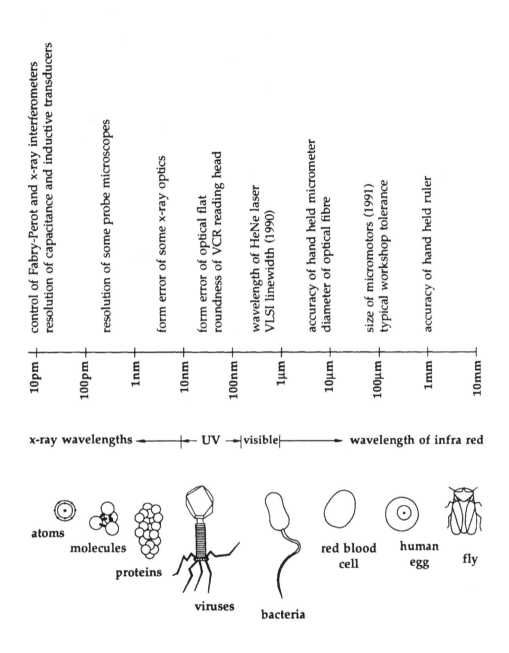

Figure 1.1 Scale bar showing the relative sizes of many familiar and unfamiliar systems

Except for a few special cases, it is difficult to envisage conventional approaches coping with such levels of precision, yet, as pointed out by Richard Feynman, 1961, there is still plenty of room at the bottom. Perhaps it is better to abandon traditional engineering methods and return to its roots in the basic models of physics, designing and manipulating from the atomic level upwards. This is the approach identified by some authorities as that of 'Nanotechnology', although we prefer the wider view that it should contain the best of both approaches in successfully addressing precisions in the 0.1 to 100 nm region, Taniguchi, 1974. Whichever definition is preferred, the mechanical aspects of such endeavours are the clear domain of ultra-precision engineering.

Where exactly lies the limit of precision for engineering applications? If we consider that an artefact must be produced, then the atomic dimension provides a limit on physical size and dimensional tolerance. However, to manipulate at the atomic level requires measurement and control of displacements to between a tenth and a hundredth of that limit. As will be seen, there are some jobs that need even crazier displacement sensitivities. The crystal radius of the carbon C^{4+} ion is around 0.15 nm, Pauling, 1945. Bonding together a few carbon, oxygen, hydrogen and nitrogen atoms results in amino acid molecules that are around a nanometre in length and which combine to form the basic proteins. In this region of size we find the tolerances for x-ray optics and the layer structures of electronic and optoelectronic devices. Increasing in dimension by an order of magnitude we are back at the scale of virus structures, (which some consider as equivalent to nanotechnological machines), and at the tolerance to which video recording heads and many other practical engineering components must be manufactured. Also around this level brittle materials often exhibit ductile behaviour; boundary lubrication film thickness is normally to be found; errors in diffraction gratings and optical cavity interferometers can seriously affect performance; and surface roughness begins to affect the reflectivity of a surface. Another order of magnitude increase to a few hundred nanometres sees conventional manufacturing techniques becoming feasible, Franse, 1990.

At the (sub-) nanometre level mechanical operations such as moving, machining and joining will be indistinguishable from the processes of chemistry and biology. The excitement of nanotechnology is that it is a unique engineering discipline that promises to have applications in all areas of science and industry. Engineering tools are rapidly being developed to fulfil the desire to operate routinely at this level. Seminal to their development was the invention of probe microscopy by Russell Young in 1972. His concept has since developed into a more generalized field of probe microscopy, in particular by Binnig *et al.*, 1982 and 1986. These techniques regularly demonstrate subnanometre measurement resolution and under specific conditions are capable of achieving atomic scale surface modifications, Miller and Hocken,

1990. Several types of sensor and actuator used in engineering instrumentation appear capable of resolutions best expressed in picometres. Some complete instrument structures, and by no means all of them esoteric research tools, need sub-atomic positional stability. Two such are Fabry-Perot etalons and x-ray interferometers – both of which require control at the picometre level if they are to operate as well as is desired, Hicks et al., 1984, Bowen et al., 1990.

Ultraprecision engineering includes the discipline of scientific instrument making and it is from this that many of our examples are taken. This book is concerned with the design of mechanical mechanisms that can operate with tolerances within the nanometre or sub-nanometre regime. The same approaches may often profitably be used when somewhat less stringent tolerances will suffice.

There is a balance between analysis and explanations of why designs have evolved in the way that they have so that past lessons may be related to future designs. The techniques to be explored are relevant not only to the direct control of microdisplacements but also to any field where dimensional stability is needed; machines that incorporate the likes of force actuators, microdevices, balances and various probes in which nanometre level distortions or disturbances may lead to significant errors.

This book discusses both instruments and machines. The former is considered to be a device or system that is designed to monitor the state of a process without influencing its behaviour while a machine is a device that is intended to operate on a system in order to cause a controllable change.

1.2 Instrumentation terms

It is with some regret that we feel it necessary to break off from the main theme of the book in order to review definitions and, in the next section, errors. Some terms in common use by metrologists may be less familiar to others and in any case familiarity does not ensure precise use. We know of many occasions when badly phrased descriptions have led to inferior instrument capability or the faulty interpretation of results. A fuller discussion of the implications of such terms as accuracy and precision as they relate to the expression of measurement quality is given by Hayward, 1977.

Range As its name implies, this is the extent over which the instrument or machine can reliably function within the confines of its specification.

Resolution or discrimination This the smallest discernible change in the parameter of interest that can be registered by a particular instrument.

Repeatability Repeatability is a measure of the scatter of results obtained if an attempt is made to exactly repeat a given operation. There is no uniquely agreed measure, but commonly either the standard deviation or the half-width of the distribution of the results might be used. Unless one is prepared to truncate values to a numeric precision much smaller than the system capability, repeated measurements are doomed to produce varied results because errors are always present and are fickle by their very nature. The repeatability of a system is not related to its linearity except when non-linearities are large enough to cause a multi-valued (non-monotonic) characteristic. Poor mechanical repeatability is commonly associated with hysteresis due to backlash or is simply inherent to the transducer as, for example, with piezoelectric gauges. With careful design, the repeatibility may approach, but can never be better than, the resolution.

Accuracy Strictly, accuracy is the deviation of the measured value from the 'true' value. This is not actually very helpful, since it is not determinable. More practically, the accuracy of an instrument is taken as a statement of the expected (or worst case) deviation of a measured result from the true value, usually expressed as a fraction either of that value or of the full scale range of the system. 'Accuracy' and 'precision' should be rigourously differentiated. The latter, associated with repeatability, is necessary but not sufficient for the provision of an accurate device. The precision with which an instrument can be used would be unaffected by a mis-calibration of 20%, but its accuracy certainly is.

Error The amount by which an assumed value deviates from its true value, error is closely associated with accuracy. Errors belong to two distinct classes. Random errors are related to the precision, or scatter, of results and may be treated statistically: for example, averaging may be used to improve the precision of a best estimate value. The magnitude of such errors may be judged from the results of a set of repeat measurements. Systematic errors are those that occur in the same way at every measurement and so cannot be discovered merely by examination of the results. Mis-calibration is a simple example. The search for, and reduction of, systematic errors is both a major chore and an ultimate source of inspiration for the instrument designer. Systematic errors are also the main reason why it is much more expensive to provide good accuracy than merely good precision.

Precision 'Precision' is used with several different meanings in engineering contexts. Sometimes it is a synonym for repeatability, as defined above. Often it is regarded as the ratio of resolution to range (occasionally, of accuracy to range). A third usage takes a precision system to be one that provides better accuracy or smaller resolution than is typically obtained. There is a mixture of the relative and the absolute in these definitions. Also each tends to imply the preceding ones in practice, although a device working at a technological limit might be considered precise even if it has a very limited range. Often precision can be imposed on a system through just one element of a closed loop process. For example, in digging the tunnels, roughly 25 km long, under the English Channel the ends aligned to within a fraction of a metre. The range to accuracy of approximately 100,000:1 did not require civil engineering machinery of high precision. It was sufficient to provide precision optics in the laser alignment and other surveying tools that were used to guide the tunnel drilling machines. Similar methods ensure the extremely precise alignment of the tubes of the Stanford Linear Accelerator, Jones, 1968. In general use the phrase 'precision engineering' tends to be given a mechanical engineering context and also an absolute resolution measured in fractions of millimetres or less. This may still encompass objects ranging in size from large structures such as aircraft down to microactuators measuring tens of micrometres or less.

Traceability The interchangeability of engineering components and most of the needs of 'legal metrology' depend on there being internationally agreed definitions of measurement units. These are of no use on their own and there must be also an established method of relating practical measurements to the Standards that embody those definitions. Any real measurement has two sources of errors. There are both systematic and random errors associated with the specific measuring process and a further systematic error caused by the imperfect conformance of the instrument to the Standard. Essentially this is a calibration error: the best we can do is to state that an instrument is calibrated to a higher authority device with an error not exceeding some specified amount. That calibration device will itself be set against another system to some specified accuracy and so on in a continuous chain of calibrations that ends with a direct comparison to the fundamental Standard itself. Any measurement in which all the stages of this chain are known is said to be 'traceable' to the Standard. The total possible error due to the imperfection of all the calibrations will be known. Generally, each successive stage of calibrating a calibration system becomes more precise, more expensive to undertake and less frequently performed as the Standard is approached. Normal practice is to establish a local standard that is tested against a National Standard. The local standard is not used directly since each access to it increases the chances of it being altered. Instead a transfer standard is used for calibrating working systems while itself being regularly checked against the local standard.

1.3 Errors

While 'error' has been defined above, we devote the final section of this chapter to a brief overview of some fundamental error sources and the basic analytic methods of handling them. Again, this is done to establish a background of notation and nomenclature and so the treatment makes no claims to completeness or rigour. Cases of direct significance will be dealt with as they arise throughout the book.

The objective of a measurement is to isolate and quantify a single parameter, often called the *measurand*, that describes a physical system state or phenomenon. Errors occur when the measurement includes other influences intrinsic to the design or effects of which the experimentalist is ignorant. Philosophically it is not possible to have a complete knowledge of all physics involved in a measurement and pragmatically it is not possible to design a perfect instrument that completely isolates the phenomenon to be observed. Furthermore most of the phenomena we might measure are considered to exist within a continuum and yet we must describe them by numbers of limited precision. All of these might be taken as approximations in our model of the world or as errors in our measurement according to circumstances and in either case they limit our knowledge.

Other sources of error are more directly related to the method of measurement and so more amenable to improvement. These include random effects caused by noise and inaccuracy in the datum of comparisons, which is how nearly all measurements are performed. For example, one may measure the mass of an object by balancing it against 'known' weights on an assay balance. The degree to which we can trust these is determined by their traceability. The degree to which we can trust the comparison, even if we had exact weights, depends on the precision of the levers in the balance and on the behaviour of the pivot and background vibration. The weights and levers provide systematic errors, vibration is probably reasonably treated as giving random errors and the pivot is likely to contribute to both classes.

On a practical level, a successful measurement is assumed if the observed values are consistently close to those that are predicted or expected. The tolerance that we specify as sufficient is often termed the confidence interval. There is a great danger in working relative to expected values (which will not be based on complete information) that a systematic but unexpected effect may be present and remain undetected. One of the most difficult tasks of the experimentalist is that of deciding when a measurement is 'correct' to the tolerance required. It is a fallacy to claim that it is easier to guarantee, say, 10% than 1%, although probabilistic statements about the relative likelihood of overstepping different tolerance bands may be valid.

1.3.1 Random errors and noise

All instrument systems contain noise. On a fundamental level there will be an intrinsic noise energy of magnitude $kT/2$ where k is Boltzmann's constant (1.38

x 10^{-23} J K^{-1}) and T is the absolute temperature. It is this noise that accounted for the erratic motion of pollen on water, a phenomenon first observed by, and later named after, Robert Brown in 1827 and explained from quantum viewpoint by Albert Einstein in 1906. Although attributed in this particular case to the collisions between molecules in the water and the pollen, an equivalent effect is intrinsic to all bodies. To quote from the excellent paper of Bowling-Barnes and Silverman, 1934,

> The statistical laws governing the behaviour of our system have nothing to say regarding the detailed mechanism of the energy fluctuations. It is therefore unsatisfactory to attempt in a particular case to say that the source of the existing variations is this or that effect. Furthermore we may not hope that by removing one of these supposed causes, let us say by [placing the system in a vacuum] and thus preventing bombardment by air molecules, to decrease the magnitude of these fluctuations. There must be, regardless of what the system and the cause may be, a mean kinetic energy of exactly $kT/2$ associated with each degree of freedom of the system.

For example, a spring balance of stiffness λ experiences an average positional fluctuation at room temperature of magnitude 0.06 nm divided by the square root of the stiffness, Bowling-Barnes and Silverman, 1934. Although this is normally totally negligible, it might become significant in micromechanical devices having low stiffness or other applications having a very high displacement sensitivity.

Usually, thermally excited mechanical noise can be ignored because larger sources dominate, for example; external vibrations transmitted through the air and through foundations; internal vibrations originating from out of balance forces of moving components and turbulent flow of fluids and gases; wear of components leading to erratic behaviour; the rubbing of surfaces having roughness on the microscopic scale; and electrical noise, perhaps itself of thermal origin, in sensors and actuators. The vibration of mechanical systems is discussed in Chapter 10.

For present purposes we assume that the noise cannot be reduced at source and so concentrate only on a review of basic techniques for extracting the best quality information from instrument signals. Consider an experiment to determine a fixed value of a system property. Assuming that the value remains constant during the test and that there are no systematic errors in the instrument (a rather rash assumption) then the instrument signal will be the true value plus a time varying noise. If we denote the output at time t by $x(t)$, the true value by μ and the noise by $n(t)$ then

$$x(t) = \mu + n(t) \tag{1.1}$$

The practical consequence is that the accuracy to which we can hope to determine μ is governed by the magnitude and time behaviour of the noise. The assumption that the noise is independent of the measurement, which is itself independent of time, is generally made in order to reduce the complexity of analysis required for an understanding of the measurement (for a discussion of other cases see statistical texts such as Yule and Kendall, 1945, or Cramer, 1946). To further reduce the complexity of signal analysis we also assume that; the noise signal will maintain its general characteristic over time, or be *stationary*; the individual samples of the signal will also, on average, show similar characteristics, or be *ergodic*; and the noise is evenly distributed about the true value. A typical signal from such a process is shown in Figure 1.2. This signal could represent a result as sampled by a digital computer. Although plotted as a continuous signal it is in fact a series of numbers sampled from the real signal and plotted in a time sequence. Consequently, all information about behaviour between samples is lost, itself a potential source of error or noise. Plotted alongside the signal is an indication of the relative amounts of time spent at different amplitude levels. The same information is given in the lower part of Figure 1.2 in the more familiar form of a frequency or probability density function.

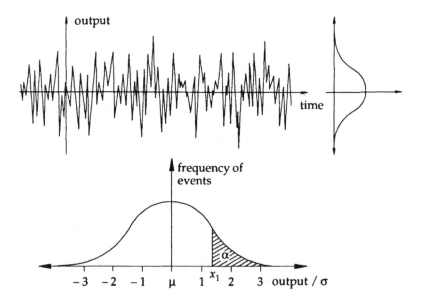

Figure 1.2 A typical output signal containing random noise, a) signal, b) distribution of noise about the mean value normalised to the standard deviation of the output, σ

The best estimate we can obtain for the true value will be some form of average of this distribution, perhaps the most likely value, or mode. When the noise is evenly distributed with decaying amplitude, and in many other real situations, the best estimate (in a statistical sense) of the true value is the *mean*, which will be the same as the mode if the symmetry is complete. Its true value can only be found if a very large number of samples (theoretically infinite) are taken. In reality there is neither the time nor computational facilities for such a measurement and estimates based on a finite set of samples must be used. Commonly the decay envelope of the distribution (or histogram if we use a small set) follows fairly closely the exponential square law characteristic of a normal or Gaussian distribution of the noise. In this instance there are a number of statistical methods for dealing with the data. For any sample of N values of which the ith is denoted x_i the mean of the sample, $\bar{x}$ is

$$\bar{x} = \frac{1}{N} \sum_{i=1}^{N} x_i \qquad (1.2)$$

For calculations of more sophistication than this we must normally assume that each value is *independent* of the previous history.

The breadth of the distribution about the mean value is expressed by the population variance, σ^2, or standard deviation, σ. Variance is the mean square deviation of ordinates from the mean. If we have only a set of N points from which to determine the distribution statistics, the best estimate is the sample variance, s^2, defined as

$$s^2 = \frac{1}{N-1} \sum_{i=1}^{N} (x_i - \bar{x})^2 \qquad (1.3)$$

Dividing simply by N would give a biassed result: the calculation involves the mean which has already been calculated from the same data set. For larger samples, say of 100 or more, the sample variance becomes nearly the same as the population variance and the distinction is unimportant. In small samples it can be critical.

Commonly distributions such as Figure 1.2 are normalized so that both the total area enclosed is unity (essential if it is to be a probability density) and the horizontal axis is scaled in units of the standard deviation centred at zero. The area, α, enclosed between the curve and any region of x values is then the probability that a single measurement, x_i, will be in that region. These area integrations can be found tabulated in any statistics text, see for example Kennedy and Neville, 1986. A brief example of the approach usually adopted is shown in Table 1.1 which expresses a probability, α, in terms of the normal variable $u = (x - \bar{x})/\sigma$. It gives the probability that a measured value from a normally distributed population will lie between u and infinity. Therefore, because the distribution is symmetrical, we can say that there is a 0.26% chance that a sample x_i will be either less than or greater than 3σ or, alternatively, there is a 99.74% probability that the measurement will be between $\pm 3\sigma$. The statistical approach allows a quantitative degree of confidence to be attached to a measurement. We may say that the best estimate of the true value of a measurand is $\bar{x} \pm 3\sigma$ to 99.7% confidence or $\bar{x} \pm 2\sigma$ with about 95% confidence.

u	0.5	1.0	1.5	2.0	2.5	3.0	3.5
α (%)	31	16	8.7	2.3	0.26	0.13	0.02

Table 1.1: Areas under the normal curve

Particularly if only a small sample is taken from a system with random noise, the sample mean may vary significantly from the actual mean value. Fortunately the random variation of the sample mean will in itself be normally distributed with a standard deviation $S_{\bar{x}}$ given by

$$S_{\bar{x}} = \frac{s}{N^{1/2}} \qquad (1.4)$$

The closeness of the estimated value to the actual mean increases in inverse proportion to the square root of the number of samples. Again, because it is normally distributed, a confidence can be attached to the mean value based on the sample variance.

Real distributions will not be exactly normal and many take other distinct forms. For example, event counting such as the number of x-ray photons arriving at a detector tends to follow a Poisson distribution. In practice normal statistics are quite robust and the simple methods outlined work well even with readily noticed deviations from the shape of the normal curve. In particular, we take advantage of the *Central Limit Theorem* which states that for any random variable the distribution of sample mean values defined by Equation 1.2 will tend to be normal about the population mean with a variance given by Equation 1.4.

We end by noting that there are severe practical restrictions on the extent to which statistical methods can be used to 'improve' the accuracy of measurement. From the above analysis it might appear that a rapid and accurate estimate of the true mean can be obtained by taking a large number of samples at a fast rate. However, this is only valid if the signal is changing at a rate sufficient to ensure that each sample is totally independent of the previous one. For a slowly changing signal the sampling rate must be reduced so that enough time elapses for the measured value to have changed sufficiently to make the samples effectively independent. It is possible at most to sample at the speed of the noise for statistical techniques to remain valid. This places a constraint on the speed with which accuracy can be achieved regardless of the size and power of the computer that may be used for sampling.

1.3.2 Error combination

Systematic errors can be sub-divided into two important categories. Some are 'correctable errors' typically caused by manufacturing tolerances, calibration error, and linear effects such as elastic distortion or thermal expansion that, once identified, repeat so precisely that a compensation for them can be built into a measurement system. For example, pitch error in screw threads posed problems for early precision engineers. Non-uniform lead screws produce a jerky motion of the nut when the thread is rotated at a uniform rate. If these are then used in lathes for producing more threads the error will propagate and a systematic worsening occur. However, the error is very repeatable and can be compensated by attaching to the nut an arm which is forced to follow a profiled tangential path. Of course, the appropriate profile must be established experimentally during calibration.

More commonly, systematic errors will be unpredictable in the sense that, although they repeat within a specific set of measurements, their long term variability is sufficient to prevent simple compensation. The obvious example is ageing effects. Other common sources are include hysteresis, backlash, thermal or temporal drift, creep, wear and inelastic distortion in mechanisms. They are the reason that periodic recalibration of instrumentation is necessary.

A completely different error that is often best considered as unpredictably systematic occurs when a result is obtained through the combination of a series of measurements. For example, if the volume of a cube was determined by three orthogonal measurements between its faces, we would expect to accumulate errors from each individual measurement so that the overall uncertainty of its volume is greater than that for any of the lengths. As a more illustrative example of the effects of cummulative measurements, consider an experiment in which a force applied at the end of a cantilever beam is to be determined by measuring its deflection δ. As shown in Chapter 2, the relationship between the force, F, and the displacement of the beam is

$$F = \frac{Ebd^3\delta}{4L^3} \qquad (1.5)$$

where b, d and L are the beam dimensions and E is the material stiffness.

The force is a function of the five measured values i.e. $F = f(E,b,d,\delta,L)$. Differentiating F with respect to all five variables by using the chain rule of classical calculus gives

$$dF = \frac{\partial F}{\partial E}\,dE + \frac{\partial F}{\partial b}\,db + \frac{\partial F}{\partial d}\,dd + \frac{\partial F}{\partial \delta}\,d\delta + \frac{\partial F}{\partial L}\,dL \qquad (1.6)$$

Substituting Equation 1.5 into 1.6 and dividing through by F gives the fractional change in force with each of the five measured variables

$$\frac{dF}{F} = \frac{dE}{E} + \frac{db}{b} + 3\frac{dd}{d} + \frac{d\delta}{\delta} - 3\frac{dL}{L} \qquad (1.7)$$

Each of the ratios in this equation indicate the tolerance within which values are known relative to their overall value. Errors in measurands that are multiplied together to derive a result have an additive effect on the overall tolerance. Also raising a variable to a power contributes a proportionate scaling of its error to the overall value. The negative sign in the last term arises because of the inverse cubed relationship of the beam length. This indicates that, for instance, if we believe that the thickness of the beam is likely to be overestimated, it would be best to err to the smaller side for the length. In practice we must never depend on such lucky coincidences, and since we cannot be sure of the trend in any error term, the overall uncertainty is obtained by simply *adding the magnitude* of all terms. If a result is derived from the addition of measurands, then their individual actual errors, not their fractional ones, add.

If the errors are systematic we should probably work with worst case conditions in Equation 1.7. Some feel that this is pessimistic since we would not expect everything to work for the worst if the sources are independent. Hence it is quite a common practice to treat the combination of systematic errors in the same way as may be done for random ones. Random errors have no theoretical worst case and it would clearly be unduly pessimistic to expect that independent random events are coincidentally highly correlated, so we seek an overall variance. Statistically, we assume that the errors are normally distributed. It can then be shown that if each parameter is independent and has a known variance, σ, the total variance of the resulting force, from Equation 1.5, about its true value will be

$$\frac{\sigma_F^2}{F^2} = \frac{\sigma_E^2}{E^2} + \frac{\sigma_b^2}{b^2} + \frac{3^2\sigma_d^2}{d^2} + \frac{\sigma_\delta^2}{\delta^2} + \frac{3^2\sigma_L^2}{L^2} \qquad (1.8)$$

The total variance is the sum of variances of all the constituent variables multiplied by their respective power law relationship to the measurand. Squaring all terms has eliminated the negative sign from this expression. It is normally acceptable for combinations of parameters that are additive to directly add variances. By analogy, (root) mean square estimates are used also for non-random error combination.

These examples illustrate a general law of design that the more complex the instrument system, the more potentially erroneous the result.

References

Binnig G. and Rohrer H., 1982, Scanning tunneling microscopy, *Helv. Phys. Acta*, **55**, 726-735.

Binnig G., Quate C.F. and Gerber Ch., 1986, Atomic force microscope, *Phys. Rev. Letts.*, **56**(9), 930-933.

Bowen D.K., Chetwynd D.G. and Schwarzenberger D.R., 1990, Sub-nanometre displacements calibration using x-ray interferometry, *Meas. Sci. Technol.*, **1**, 107-119.

Bowling-Barnes R. and Silverman S., 1934, Brownian motion as a natural limit to all measuring processes, *Rev. Mod. Phys.*, **6**, 162-192.

Cramer H., 1946, *Mathematical Methods of Statistics*, Princeton University Press, Princeton N.J.

Feynman R.P., 1961, There's plenty of room at the bottom, in *Miniaturization*, ed. by H.D. Gilbert, Reinhold Publishing Corp., New York, 282-296.

Franse J., 1990, Manufacturing techniques for complex shapes with submicron accuracy, *Rep. Prog. Phys.*, **53**, 1049-1094.

Hayward A.T.J., 1977, *Repeatability and Accuracy*, Mechanical Engineering Publications, London.

Hicks T. R., Reay N. K. and Atherton P. D., 1984, The application of capacitance micrometry to the control of Fabry-Perot etalons, *J. Phys. E: Sci. Instrum.*, **17**, 49-55.

Hume K.J., 1980, *A History of Engineering Metrology*, Mechanical Engineering Publications, London.

Jones R.V., 1968, More and more about less and less, *Proc. Roy. Inst. Gr. Brit.*, **43**, 323-345.

Kennedy J.B. and Neville A.M., 1986, *Basic Statistical Methods for Engineers and Scientists*, Harper and Row, New York.

Miller J.A. and Hocken R.J., 1990, Scanning tunneling microscopy bit making on highly oriented pyrolytic graphite: Initial results, *J. Appl. Phys.*, **69**(2), 905-907.

Pauling L., 1945, *On The Nature of the Chemical Bond*, Cornell University Press, Ithaca NY.

Taniguchi N., 1974, On the basic concept of nanotechnology, *Proc. Int. Prod. Eng. Tokyo* (Tokyo JSPE) Part 2, 18-23.

Young R., Ward J. and Scire F., 1972, The Topographiner: An instrument for measuring surface microtopography, *Rev. Sci. Instrum.*, **43**(7), 999-1011.

Yule G.U. and Kendall M.G., 1945, *Elements of Statistics*, Griffin, London.

2 INTRODUCTORY MECHANICS

The design of instrument mechanisms should be based firmly upon an understanding of both static and dynamic forces. We exploit their action on rigid bodies and on elastic systems where there is intentional subsequent distortion. A brief review of introductory mechanics models is therefore included here to establish the analytic techniques that are used regularly in this book. The key aspect of the statics section is simple elasticity and beam theory. Because the majority of instrument structures are relatively rigid, dynamic analysis concentrates on energy methods and, especially, on Lagrange's Equation. Although seeming laborious for very simple systems, we advocate this approach because it is highly systematic and may readily be applied to complex dynamical systems. The dynamics of distributed bodies are considered through the vibration of beams, including Rayleigh's method for the simple estimation of fundamental frequencies of vibration.

Introduction

It is far from our minds to attempt to cover the vast field and subtleties of mechanics in one chapter. Our aim is merely to present in one place the background to some of the formulae that are used sporadically throughout the book so that design ideas are not hidden under detail in later chapters. For brevity, given its purpose, formal proofs are rarely included and the reader wanting a more comprehensive treatment is referred to texts such as Ryder, 1969, Timoshenko and Young, 1968 and Arya, 1990. Following standard convention, mechanics has been divided into static and dynamic systems. The essential difference between them is that in statics we assume that accelerations can be ignored: note from the outset that a body in uniform motion can be treated by the methods of statics.

The modelling of almost all mechanical systems starts with some or all of four basic abstractions; the particle, the rigid body, the pin joint and the rigid joint. The particle is a point mass that requires coordinates to describe its position within a frame of reference. The rigid body has extension as well as mass and so requires additional coordinates to describe its orientation relative to the reference frame. A rigid body maintains its exact shape irrespective of the forces that it carries. A pin joint is a connection between bodies that allows relative rotation without any transmission of torque or moment. A rigid joint is one that allows no relative motion and is distinguishable only in marking the notional boundary between a rigid body and one that is allowed to show elastic behaviour. Considering how far all these ideals are from what is seen in real systems, it is somewhat remarkable that models using them are often perfectly adequate for design and analysis.

2.1   Statics, elasticity and beam distortion

Since statics is concerned with systems having negligible acceleration, it is also essentially concerned with bodies, or systems of bodies, that are in force equilibrium. It is convenient first to look at simple frames and the basic definitions of elasticity and then to treat separately the important case of beams subject to bending.

Load bearing structures are usually analyzed as either pin-jointed or rigid frames. The advantage of the former lies in the fact that if no moment can pass across a joint, then there can be no moments anywhere in a member that is not subject to transverse loads. The pin-jointed frame is loaded *only at its joints* so that members carry only simple tensile or compressive loads. By suitable geometrical arrangements, the whole of a structure can be solved by examining successively the force equilibrium at each joint. If joints are found at which the forces cannot be balanced the system is a mechanism, that is, it is underconstrained. If there are too many members for a unique resolution of forces to be found, then it is statically indeterminate. The issues of constraint will be further explored in Chapter 3. Here we note that the statically determinate geometry is the one familiar from truss structures on roofs and bridges; members in interconnecting triangles, for a planar structure, or tetrahedra. Note also from the observation of bridges that bolted joints are treated as pin joints without undue error. Only the slightest ability to rotate is needed to make the assumption a reasonable first approximation. Another implication of insisting that all load is carried at the joints is that the self weight of the members must be small compared to the overall loading.

When there are significant transverse loads on members of a structure, other methods must be used. If the joints can be taken as pins, it may be sufficient to treat the load as shared by the ends of the member in order to solve the overall system and then to return to each member so loaded and treat it as a simply supported beam under combined axial and transverse loads. Some special pin-jointed structures, such as the three-pin arch, have simple formal solutions. If the joints are rigid, the system either acts as a rigid body or there is elastic distortion of the members relative to each other. In this case, as always when the system is statically indeterminate, we use elasticity theory to predict shape changes and apply continuity and compatibility conditions at the joints to provide the extra equations we need. These amount to requiring that the whole structure remains intact and that any small part is in equilibrium from the stresses around it. For the mechanisms and structures regularly met in this book, the members are often quite long compared to their cross-sectional dimensions and are relatively straight. It is reasonable to assume that distortion through shear is very small compared to that from bending. This is the basic simplification of beam theory, which is the main topic here.

2.1.1 Elementary elasticity

All real objects will distort under an applied load. The magnitude of the distortion and the stress imposed on individual members can introduce unwanted motion and possibly even failure of the structure. For the present we take it that no gross motion occurs and that the material behaves elastically. Except if explicitly stated, we assume throughout this book that materials are both homogeneous and isotropic. This is definitely imprecise but is normally adequate for common structural materials such as metals. It is inadequate, for example, for composites or single crystals. The engineering definition of direct stress, σ, in tension or compression is the ratio of a uniform normal load, F, divided by cross-sectional area, A, see Figure 2.1a

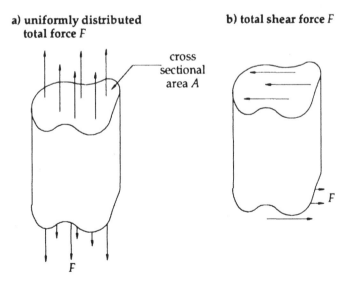

Figure 2.1 Direct stresses and shear stresses

$$\sigma = \frac{F}{A} \qquad (2.1)$$

From Figure 2.1b the shear stress is defined as the component parallel to the load, F, caused by the action of a pair of opposed but non-aligned forces, and has the value

$$\tau = \frac{F}{A} \qquad (2.2)$$

Under modest load conditions, the displacement per unit length, or direct strain, ε, is proportional to the stress through a constant, E, known as the elastic modulus or Young's modulus. A thin, uniform prismatic rod of length L subject to a force that produces an axial displacement δ is strained by

$$\varepsilon = \frac{\delta}{L} = \frac{\sigma}{E} \qquad (2.3)$$

Similarly, the angular distortion or shear strain, γ, of a body is proportional to the shear stress, τ, through a constant known as the modulus of rigidity, G,

$$\tau = G\gamma \qquad (2.4)$$

Stress does not usually occur along a single linear axis but has a three dimensional character. It can be resolved into three orthogonal components, called the principal stresses, along a set of principal axes which carry zero shear stress. The maximum shear stresses occur along three orthogonal axes which are at 45° to the principal stresses and each has a value one half of the difference between the two principal stresses in whose plane it is situated, for example

$$\tau_{xy} = \frac{(\sigma_x - \sigma_y)}{2} \qquad (2.5)$$

The strain along a principal axis is influenced by stresses in other axes by another constant of proportionality known as Poisson's ratio, v, so that

$$\varepsilon_x = \frac{(\sigma_x - v(\sigma_y + \sigma_z))}{E} \qquad (2.6)$$

The second two terms of Equation 2.6 represent the additional, negative, strain in the x axis due to direct stresses in the in the other two axes. At its simplest this merely reflects the argument of continuity that if we stretch a rod it should get thinner. With real materials there is a mixture of genuine volume change and redistribution, so Poisson's ratios are material specific and in non-isotropic materials the values may be different in different planes (technically, this is because stress and strain are properly treated as second order tensors, but we need not be concerned with that here). Volumetric considerations restrict the value of Poisson's ratio to below 0.5, although it can go to zero, and even be negative, in specific directions in some materials.

Because of the interrelated nature of the shear and direct stresses, it can be shown that the modulus of rigidity and the elastic modulus are related by

$$E = 2G(1 + v) \tag{2.7}$$

2.1.2 Bending of symmetrical beams

The elastic distortion of structures under an applied force may be a problem to be overcome in a design process. It is also commonly used for creating smooth, repeatable motion of very high precision. The stress induced by this force must not exceed the yield stress in a ductile material or the breaking stress in a brittle one, else the system behaviour will be permanently modified! In practice, it is very commonly found that transverse (bending) behaviour dominates over the axial. It is normally adequate to compute both the deflections and the stress levels in the presence of transverse forces by simple beam theory. Consider the beam element ABCD, of length dx, shown in Figure 2.2a, part of a 'long, thin' prismatic member that is subject to bending. If there is a small, uniform bending moment, M, about the z axis (going into of the paper), the beam will distort into an arc having a large radius of curvature, R. Selecting a slice, aa, in the xz plane it is clear that the distortion will induce a strain ε_x given by, Figure 2.2b,

$$\varepsilon_x = \frac{(R + y)\delta\theta - R\delta\theta}{R\delta\theta} = \frac{y}{R} \tag{2.8}$$

where it is assumed that the central, *neutral axis* remains its original length, so $dx = R\delta\theta$, and that plane sections such as AB remain plane. Substituting Equation 2.3 into 2.8 and rearranging gives

$$\frac{\sigma}{y} = \frac{E}{R} \tag{2.9}$$

Because this direct stress is caused by a pure bending moment, the total normal force on the beam is zero. Therefore we can write, using integration over planes such as CD,

$$\int \sigma dA = \frac{E}{R} \int y dA = 0 \tag{2.10}$$

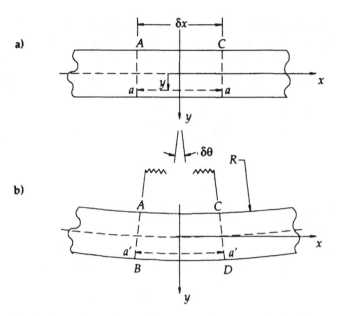

Figure 2.2 Deformation of a prismatic beam element subject to a bending moment about the z-axis

where dA is an elemental strip of the cross-section in the z direction. The integral on the right indicates that the first moment of area about the chosen axis must be zero. Consequently, the neutral axis is coincident with the centroid of the beam. Also, the total bending moment in the beam is given by

$$M = \int \sigma y dA = \frac{E}{R} \int y^2 dA = \frac{EI}{R} \tag{2.11}$$

where I is the second moment of area of the beam about the neutral axis. Combining Equations 2.9 and 2.11 gives the general equations for the bending of a beam about its neutral axis

$$\frac{M}{I} = \frac{\sigma}{y} = \frac{E}{R} \tag{2.12}$$

Note also that by similar arguments it can be shown that the equations governing pure torsion about the centroidal axis of a rod are given by

$$\frac{T}{J} = \frac{\tau}{r} = \frac{G\varphi}{L} \tag{2.13}$$

where r is the radial distance from the axis and J is the polar second moment of area $(= \pi r^4/2$ for a rod of circular section).

The radius of curvature of a line in the xy plane is given by

$$\frac{1}{R} = -\frac{\dfrac{\mathrm{d}^2 y}{\mathrm{d}x^2}}{\left(1 + \left(\dfrac{\mathrm{d}y}{\mathrm{d}x}\right)^2\right)^{\frac{3}{2}}} \qquad (2.14)$$

where the negative sign indicates a sagitta of the beam as shown in Figure 2.2.

For an originally straight beam under the influence of relatively low applied moments, the slope of the distorted beam will be small and the denominator of Equation 2.14 can be taken as unity. In practice it is rare to find elastic distortions large enough to make this unreasonable. Then introducing Equation 2.12 gives

$$EI\,\frac{\mathrm{d}^2 y}{\mathrm{d}x^2} = -M \qquad (2.15)$$

An expression for the deflection, and consequently the stiffness, of a beam can be derived through two successive integrations of Equation 2.15, provided that an expression for the bending moment of the beam and the boundary conditions are known. This method is used explicitly in Section 4.1 and so need not be further illustrated here.

Under certain circumstances where only the deflections of a beam under a pure bending moment are required, it is more convenient to use the moment area method. This method of solution is commonly employed by civil engineers for *statically indeterminate* beams and has consequently tended to be reserved for them. For a beam of length L, Equation 2.15 can be rearranged and integrated to give

$$-\int_0^L \frac{M}{EI}\,\mathrm{d}x = \int_0^L \left(\frac{\mathrm{d}^2 y}{\mathrm{d}x^2}\right)\mathrm{d}x = \left[\frac{\mathrm{d}y}{\mathrm{d}x}\right]_0^L \qquad (2.16)$$

Equation 2.16 indicates that the change in slope between two positions along the beam (from one end to the other as written here) is simply the area of the its bending moment diagram between the points (taken as negative for the sign convention used) scaled by its flexural rigidity, EI.

Similarly, multiplying Equation 2.15 by the position at any point along the beam and integrating by parts gives

$$Y_L\left(\frac{dy}{dx}\right)_L - Y_0\left(\frac{dy}{dx}\right)_0 - (Y_L - Y_0) = -\int_0^L \frac{M}{EI} x dx \qquad (2.17)$$

This equation is the basis of graphical solutions for beams of non-uniform cross-section. It provides analytic solutions if an expression for M/EI can be derived in terms of y. As an illustration, consider the beam of Figure 2.3, which is rigidly clamped between two platforms that are constrained to remain parallel but have relative movement perpendicular to the axial direction of the undistorted beam. This type of constrained beam will be further studied in Chapter 4, but some observations are appropriate here. Under deflection, end moments must be generated to maintain the parallelism and it is readily seen that the bending moments at each end of the beam are equal and opposite. Thus, as there are no loads applied directly onto the beam, its bending moment diagram will be as shown in Figure 2.3. Clearly the slope of the beam is zero at each end. Choosing the origin to be at the left support and taking the moment distribution to be $M(x) = (2M_e x/L) - M_e$, where M_e is the end moment, then Equation 2.17 gives the deflection as

$$\delta = -\int_0^L \frac{M}{EI} x dx = \frac{M_e L^2}{6EI} \qquad (2.18)$$

As more complex arrangements of beams and load conditions are introduced, the use of more advanced structural analysis methods will be required. We leave their exposition to other texts.

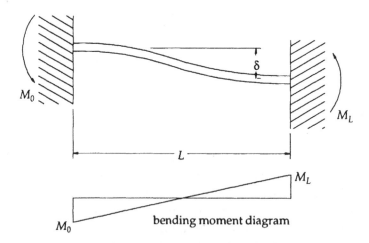

Figure 2.3   Deflection of a beam rigidly fixed between two sliding supports

2.2 Rigid body dynamics and Lagrange's Equation

Dynamics is concerned with systems under acceleration and so with forces related to inertia. If the mass of a body is concentrated into a small volume so that the forces needed to alter its orientation are small compared to those concerned with its translation, it may be treated as a particle, that is, as if all its mass exists in a single point at its centre of mass. The three laws of Isaac Newton provided the first reasonably complete model of particle motion and are still adequate for the vast majority of applications. They can be summarized as:

1. In the absence of any net force, a body will maintain a straight line motion with uniform velocity.

2. A resultant force on a body will induce an acceleration, proportional to its mass, in the direction of that force.

3. In every interaction between two bodies there must be equal and opposite actions on each of them.

The first law is the equilibrium condition, rest being a special case of uniform motion, and also establishes that Euclidean geometry is to be used to describe the motion. The second law can be stated mathematically as

$$F = \frac{d(mv)}{dt} = ma \tag{2.19}$$

The bold variables indicate vector quantities. Also, whereas position and velocity are always expressed vectorially relative to a local datum origin, acceleration must be expressed with respect to a hypothetical inertial frame of reference. In practice this restriction is not too demanding and requires only that all external accelerations of magnitudes approaching those in the system under investigation should be included. For most mechanism analysis, the Earth may be taken as stationary without noticeable error.

Throughout this book masses may be considered to remain constant, so the variable m can be moved in and out of the differential operator with impunity. The quantity in parentheses in Equation 2.19 is known as the momentum, **p**. If there are no applied forces then the first derivative of momentum with respect to time must be zero: the momentum is constant. For a constant mass, this implies a constant linear velocity and so provides a restatement of Newton's first law. More importantly it shows that for any system of bodies that is not subjected to an overall external resultant force, the total momentum must remain constant however it be redistributed between those bodies. This law of conservation of momentum appears to be one of the most rigorously exact forms yet discovered, and may be taken to over-rule any other modelling predictions with which it is in conflict.

The use of the term 'bodies' in Newton's laws must be treated carefully. Their original applications were to astronomy and ballistics, where bodies can be well treated as particles and so the more rigorous use of the latter term is often dropped. Newton, despite considerable effort, never succeeded in extending his laws to describe rigid body motion. It fell to Leonhard Euler to generalize the concept of momentum for this purpose. He provided two laws that incorporate Newton's and operate simultaneously on any body:

1. The rate of change of linear momentum of a body treated as a particle at its centre of mass is proportional to the resultant force on that body and governed by the total mass.

2. The rate of change of moment of momentum of a body about its centre of mass is proportional to the resultant moment about that centre on the body and governed by the second moment of mass of the body about its centre of mass.

The moment of momentum and second moment of mass, commonly refered to as angular momentum and moment of inertia, are minimum about the centre of mass for any body. They may be taken about other axes, but great care is then necessary in using Euler's laws consistently.

Although the more traditional vectorial methods deriving from Newton and Euler provides a direct geometrical insight, they become rather cumbersome for systems of interconnected bodies such as the mechanisms relevant to this book. In any case, the reader is assumed to have already pursued such an approach at sophomore level. Therefore in this section we take an approach first adopted in 1788 by Joseph Louis Lagrange. It is very much a *scalar* method and so readily copes with more complex systems. In reducing the vector dynamics to scalar forms it is implicitly an energy method and in using it care must be taken to ensure that the system being analyzed can be adequately described in mechanical energy terms.

The first step is quite abstract. It is not essential to formulate all equations of motion in Cartesian (or equivalent) vector spaces. A more convenient description of system behaviour may be found by choosing some other coordinates, as yet undefined. Since it moves in three dimensional space, no more than three independent coordinates will be necessary to describe the time dependent position of any *single* particle. Thus we introduce general coordinates denoted as q_1, q_2 and q_3 (see also Section 3.1). There will be a functional relationship between these and a cartesian coordinate system given by

$$x = f_1(q_1, q_2, q_3),$$

$$y = f_2(q_1, q_2, q_3), \qquad (2.20)$$

$$z = f_3(q_1, q_2, q_3)$$

Using the chain rule of differentiation, the velocity of a particle in the x direction is then

$$\dot{x} = \frac{\partial x}{\partial q_1}\dot{q}_1 + \frac{\partial x}{\partial q_2}\dot{q}_2 + \frac{\partial x}{\partial q_3}\dot{q}_3 \qquad (2.21)$$

The dot above a symbol represents total differentiation with repect to time (sometimes called the Newtonian fluxion notation). Expressions for y and z can be similarly derived. In each case the speed is an explicit function of the generalized coordinates and their respective velocities. Because the q_i coordinates are independent, differentiating Equation 2.21 shows immediately that

$$\frac{\partial \dot{x}}{\partial \dot{q}_1} = \frac{\partial x}{\partial q_1} \qquad (2.22)$$

Now consider another function, g (say) of the three generalized coordinates. Differentiation of g with respect to time yields

$$\frac{dg}{dt} = \frac{\partial g}{\partial q_1}\dot{q}_1 + \frac{\partial g}{\partial q_2}\dot{q}_2 + \frac{\partial g}{\partial q_3}\dot{q}_3 \qquad (2.23)$$

Since Equation 2.23 is general, we let $g = \partial x/\partial q_1$ to give

$$\frac{d}{dt}\left(\frac{\partial x}{\partial q_1}\right) = \frac{\partial^2 x}{\partial q_1^2}\dot{q}_1 + \frac{\partial^2 x}{\partial q_1 \partial q_2}\dot{q}_2 + \frac{\partial^2 x}{\partial q_1 \partial q_3}\dot{q}_3 = \frac{\partial \dot{x}}{\partial q_1} \qquad (2.24)$$

by direct comparison with Equation 2.21.

We now pause in this geometrical analysis to examine the motion of a particle in a conservative system. A conservative system here implies a mechanical one which does not transfer energy in the form of heat. Also, for the moment, we disallow potential fields. Consider the work done, δW_1, in a movement δq_1 of the coordinate q_1 by a force Q_1 in direction q_1 only and no changes in the other q coordinates (which are, after all, independent). There will still generally be components of motion in all three axes of the xyz system, but since work is a scalar they can simply be added to give, on relating forces and accelerations,

$$\delta W_1 = Q_1 \, \delta q_1 = m \left[\ddot{x}\delta x + \ddot{y}\delta y + \ddot{z}\delta z \right]$$

$$= m \left[\ddot{x}\frac{\delta x}{\delta q_1} + \ddot{y}\frac{\delta y}{\delta q_1} + \ddot{z}\frac{\delta z}{\delta q_1} \right] \delta q_1 \tag{2.25}$$

Taking, for brevity, just the first term of Equation 2.25, it may be rewritten as

$$\ddot{x}\frac{\partial x}{\partial q_1} = \frac{d}{dt}\left(\dot{x}\frac{\partial x}{\partial q_1} \right) - \dot{x}\frac{d}{dt}\left(\frac{\partial x}{\partial q_1} \right) \tag{2.26}$$

Substituting Equation 2.22 into the first term on the right hand side and Equation 2.24 into the second term of Equation 2.26 gives

$$\ddot{x}\frac{\partial x}{\partial q_1} = \frac{d}{dt}\left(\dot{x}\frac{\partial \dot{x}}{\partial \dot{q}_1} \right) - \dot{x}\frac{\partial \dot{x}}{\partial q_1}$$

$$= \frac{d}{dt}\left(\frac{\partial}{\partial \dot{q}_1}\left(\frac{\dot{x}^2}{2} \right) \right) - \frac{\partial}{\partial q_1}\left(\frac{\dot{x}^2}{2} \right) \tag{2.27}$$

The other two terms in Equation 2.25 can be similarly transformed. Since Equation 2.27 is scalar, the three terms in this form may simply be added to give the total force required to do work on the particle in q_1 as

$$Q_1 = \frac{d}{dt}\left(\frac{\partial T}{\partial \dot{q}_1} \right) - \frac{\partial T}{\partial q_1} \tag{2.28}$$

where T is the *kinetic energy* of the particle given by

$$T = \frac{mv^2}{2} \tag{2.29}$$

Notice that the velocity is now expressed only within a scalar and therefore Equation 2.28 gives an expression for the generalized force in one coordinate only, even though the other coordinates are included in the kinetic energy term. To obtain a complete description, the equivalent of Equation 2.28 must be derived for each coordinate in turn and the resulting set of simultaneous differential equations solved.

So far we have only considered externally applied forces yet in all but the simplest situations, *potential energy*, $U(q)$, will be present and affects the dynamics of most systems. Sources of potential in a mechanical system originate typically from strain energy in springs (= $\lambda x^2/2$ for a linear spring of stiffness λ), gravitation (= mgh for small height changes near the Earth) and electromagnetics (= $CV^2/2$ for the electrostatic field of a capacitor and $Li^2/2$ for the magnetic field of an inductor). The force, Q_u, on a particle depends upon the gradient of its potential as

$$Q_u = -\frac{\partial U(q)}{\partial q} \qquad (2.30)$$

Newton's laws imply that this extra effect may be added directly into the simpler form of Equation 2.28.

At this stage we must consider how this analysis might be applied to more complex systems involving many particles and rigid bodies held together by many springs, perhaps in a gravitational field. The answer is very simple. We just combine the equations of motion for all of the coordinates necessary to describe the position and velocity of each element. The number of such coordinates corresponds to the *degrees of freedom*, n, of the mechanism. Because all energy terms are scalar they can be added. The potentials are independent of velocity.

Hence we may define the *Lagrangian* $L = T(q_1, .. q_n : \dot{q}_1, .. \dot{q}_n) - U(q_1, .. q_n)$, which is simply the difference between the kinetic energy and potential energy of all the elements of the mechanism. Remembering the addition of Equation 2.30 to Equation 2.28, Lagrange's equations of motion for a conservative system are then

$$\sum \left\{ \frac{d}{dt}\left(\frac{\partial L}{\partial \dot{q}_i}\right) - \frac{\partial L}{\partial q_i} = Q_i \right\} \quad i = 1 \dots n \qquad (2.31)$$

The summation sign is employed in a slightly unusual way (although common to several standard texts) to indicate the set of simultaneous equations generated by using each index in turn.

The above version of Lagrange's Equation ignores energy dissipation. Real systems will include, intentionally or otherwise, a damping element that may in many cases be modelled by a retarding force proportional to velocity. This type of energy dissipation can be incorporated into Equation 2.31, as is shown in Section 10.2. However, for many of the systems covered in this book primary concern is with finding the resonant frequencies of individual systems as a measure of their dynamic response. System damping is usually low, and thus has little effect on the natural frequency and Equation 2.31 may be used safely. The subject of damping is left to the final chapter.

The application of Lagrange's Equation is best appreciated by example, although it should be stressed that examples simple enough to be easily followed can also be solved readily by other methods. The power of the method is that exactly the same procedures are followed with systems of any

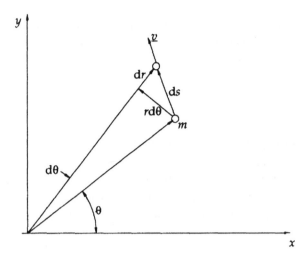

Figure 2.4   A particle moving in a plane

complexity. First consider a particle of mass m constrained to move in an xy plane, Figure 2.4. The mass is shown at two instants separated by a time interval dt. The two polar coordinates r and θ will be used, corresponding to generalized coordinates q_1 and q_2. The two velocity components of the mass can be combined by Pythagoras' theorem as

$$v^2 = (\dot{r}^2 + (r\dot{\theta})^2) \tag{2.32}$$

so the kinetic energy is given by

$$T = \frac{m(\dot{r}^2 + (r\dot{\theta})^2)}{2} \tag{2.33}$$

There is no potential term if the plane is horizontal or isolated, so Equation 2.33 is also the Lagrangian. The derivatives needed for Lagrange's equation for the two coordinates are then

$$\frac{d}{dt}\left(\frac{\partial L}{\partial \dot{r}}\right) = m\ddot{r} \; ; \; \frac{\partial L}{\partial r} = mr\dot{\theta}^2$$

$$\frac{d}{dt}\left(\frac{\partial L}{\partial \dot{\theta}}\right) = m(2r\dot{r}\dot{\theta} + r^2\ddot{\theta}) \; ; \; \frac{\partial L}{\partial \theta} = 0 \tag{2.34}$$

which on substitution into Equation 2.31 gives

$$m\ddot{r} - mr\,\dot{\theta}^2 \; = \; Q_r \tag{2.35}$$

$$2mr\dot{r}\,\dot{\theta} + mr^2\ddot{\theta} \; = \; Q_\theta \tag{2.36}$$

Equation 2.35 represents the total radial force applied to the mass, with the first term being that causing outwards linear radial acceleration and the second being the familiar centripetal force acting towards the origin. Dimensional analysis of Equation 2.36 reveals that Q_θ has the units of torque. Dividing through Equation 2.36 by r gives the tangential force at the particle that causes that torque about the origin. Then the two terms in the equation will be recognized as the familiar Coriolis and tangential accelerations respectively.

As a second example consider the case of a simple spring/mass system in a vertical gravity field* of strength g. The spring is rigidly attached at its upper end with the mass at the other. It is assumed that the mass, m, is rigid (that is in practical terms, significantly stiffer than the spring) and the spring, of stiffness λ, has negligible mass. The only coordinate is vertical displacement and the Lagrangian for this system is

$$\mathcal{L} \; = \; T - U \; = \; \frac{1}{2}m\dot{q}_1^2 - \left(\frac{1}{2}\lambda q_1^2 - mgq_1\right) \tag{2.37}$$

Substituting this into Lagrange's equation gives the familiar equation of motion for this system

$$m\ddot{q}_1 + \lambda q_1 \; = \; Q_1 + mg \tag{2.38}$$

Now, for example, if Q_1 is zero, the dynamic behaviour can be simple harmonic motion of the mass at a natural frequency $\omega_n = (\lambda/m)^{1/2}$ irrespective of the strength of the gravity field. Gravity influences only the mean extension of the spring from its natural length.

---

* The term 'gravity field' here describes the field that is apparent at a particular point on the earth, assumed constant over the distances of immediate concern. This field is dominated by the potential gradient of the Earth's *gravitational field*, but includes effects of centripetal acceleration due to its rotation and even smaller terms from its motion through space and other gravitational fields. While the differences are totally insignificant for most practical mechanism analysis, it is prudent, and no extra effort, to use a terminology that reminds us of the complex arrangement of body forces that will be experienced by a moving object.

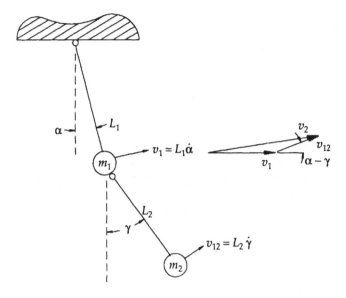

Figure 2.5   A double pendulum

As a final, more complex, system consider the double pendulum of Figure 2.5. This consists of a mass m_1 attached, by a single degree of freedom pivot, to a fixed datum via a thin inextensible rod of length L_1 with a second mass, m_2, in turn attached, by a similar pivot, to the first by another inextensible thin rod of length L_2. The whole assembly is free to swing in a vertical plane. It is most convenient to describe the motion by the two angular coordinates α and γ.

The velocity of the first mass is simply the product of the length, L_1, of the rod and its angular velocity, $\dot{\alpha}$. The absolute velocity of the second mass is less easily expressed. From the velocity diagram shown in Figure 2.5, Pythagoras' theorem in the form of the Cosine Rule shows the square of the velocity of the second mass to be

$$v_2^2 \;=\; (L_1\dot{\alpha})^2 \;+\; 2L_1 L_2\,\dot{\alpha}\dot{\gamma}\cos(\alpha - \gamma) \;+\; (L_2\dot{\gamma})^2 \tag{2.39}$$

The potential energy derives from the change in height of the masses in the gravity field and can be found using simple trigonometry. Consequently the kinetic and potential energies of the system can be expressed by the equations

$$T \;=\; \frac{m_1}{2}\,(L_1\,\dot{\alpha})^2 \;+\; \frac{m_2}{2}\left[(L_1\,\dot{\alpha})^2 \;+\; 2L_1L_2\,\dot{\alpha}\dot{\gamma}\cos(\alpha-\gamma) \;+\; (L_2\,\dot{\gamma})^2\right] \tag{2.40}$$

$$U \;=\; (m_1 + m_2)\,gL_1(1 - \cos\alpha) \;+\; m_2gL_2\,(1 - \cos\gamma) \tag{2.41}$$

These expressions could be substituted into Lagrange's equation to derive the full equations of motion. However, in many vibration problems we may assume that the magnitude of the oscillations are small perhaps, for example, because we are seeking to reduce the amplitude of resonant peaks. Under such circumstances both α and γ will be small and the cosine functions may be expanded as series and all but the most significant terms ignored to give

$$T = \frac{m_1}{2}(L_1\dot{\alpha})^2 + \frac{m_2}{2}\left[(L_1\dot{\alpha})^2 + 2L_1L_2\dot{\alpha}\dot{\gamma} + (L_2\dot{\gamma})^2\right] \tag{2.42}$$

$$U = (m_1 + m_2)gL_1\frac{\alpha^2}{2} + m_2gL_2\frac{\gamma^2}{2} \tag{2.43}$$

Assuming that the pendulum is undergoing free motion (i.e. $Q_i = 0$), substituting Equations 2.42 and 2.43 into Lagrange's equation yields the linear, homogeneous second order simultaneous differential equations

$$(m_1 + m_2)L_1^2\ddot{\alpha} + m_2L_1L_2\ddot{\gamma} + (m_1 + m_2)gL_1\alpha = 0 \tag{2.44a}$$

$$m_2(L_1L_2\ddot{\alpha} + L_2^2\ddot{\gamma}) + m_2gL_2\gamma = 0 \tag{2.44b}$$

The two natural frequencies and mode shapes of this two degree of freedom mechanism can now be solved using standard mathematical techniques, for example, through eigenvalue analysis by computer or algebraically using the method outlined in Section 10.3

A full study of Lagrangian mechanics and its extension to Hamiltonian analysis is beyond the scope of this book. The treatment provided gives sufficient insight for present purposes. For a fuller and more rigorous discussion readers are recommended to peruse the pages of Arya, 1990, Chorlton, 1985 Goldstein, 1980 and Pars, 1965.

2.3 Vibration behaviour of beams

So far we have considered only systems consisting of lumped masses and springs (with damping elements to be added in Chapter 10). However, the natural frequencies of systems with continuously distributed mass are also very important. Often continuous systems may be modelled in terms of straightforward beam elements and so it is upon them that we concentrate. Lagrange's Equation can again be used, although it becomes rather cumbersome for lateral vibrations for which a simple treatment is given. Exact solutions are analytically difficult so an approximation due to Rayleigh is also discussed. The approaches described are general and can be applied to structures other than beams.

2.3.1 Longitudinal vibration of a beam

Firstly, consider a prismatic beam of length L and mass M that is undergoing longitudinal, or axial, vibration. It is modelled by splitting it into a series of light springs and point masses of stiffness λ and mass m, see Figure 2.6. The Lagrangian for the coordinates u_i is given by

$$\mathcal{L} = \sum_{i=1}^{n} \left[\frac{m\dot{u}_i^2}{2} - \frac{\lambda}{2}(u_{i+1} - u_i)^2 \right] \qquad (2.45)$$

where u_i represents the displacement of the ith section. If the beam is uniform and each mass and stiffness represent an element of length a, their respective values are

$$m = \frac{Ma}{L} = \rho A a \; ; \; \lambda = \frac{EA}{a} \qquad (2.46)$$

where ρ is the density of the rod, A its cross-sectional area and E is its elastic modulus.

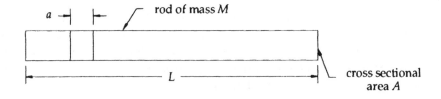

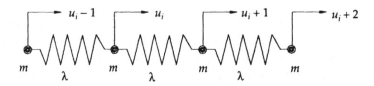

Figure 2.6 Lumped model respresenting longitudinal vibration of a prismatic rod

Substituting Equations 2.45 and 2.46 into Lagrange's Equation for free motion ($Q_i = 0$) and collecting terms containing the ith coordinate we have

$$\rho \ddot{u}_i - \frac{E}{a^2} (u_{i+1} - 2u_i + u_{i-1}) = 0 \qquad (2.47)$$

In the limit as a tends towards zero, the second term in Equation 2.47 becomes a second partial derivative along the horizontal direction, x (say), and we obtain the familiar wave equation

$$\frac{\partial^2 u}{\partial t^2} = \frac{E}{\rho} \frac{\partial^2 u}{\partial x^2} \qquad (2.48)$$

This example well illustrates the exchange of potential and kinetic energy as the fundamental mechanism of resonance. To derive the natural frequency for a specific beam, Equation 2.48 must be solved with the appropriate boundary conditions. For example, a cantilever beam of length L, rigidly fixed at one end and free at the other, would have zero extension at the fixed end (i.e. $u = 0$ at $x = 0$) and zero force, therefore also zero strain, at the free end (i.e. $\partial u / \partial x = 0$ at $x = L$). For a sustained vibration there must be a temporal standing wave in the beam, so the solution should have the general form

$$u = f(x) C_1 \sin(\omega_n t + \varphi) \qquad (2.49)$$

Substituting this into Equation 2.48 and rearranging gives

$$\frac{d^2 f}{dx^2} = -\omega_n^2 \frac{\rho}{E} f \qquad (2.50)$$

The solution for $f(x)$ is straightforward, since Equation 2.50 is just the harmonic oscillator, and so a full solution to Equation 2.48 is

$$u = C \sin \left(\omega_n \left(\frac{\rho}{E} \right) x + \beta \right) \sin (\omega_n t + \varphi) \qquad (2.51)$$

Since C cannot be zero if there is to be a non-trivial solution, the first boundary condition, that the fixed end deflection remain zero over all time, requires that the arbitrary constant β be zero. Then differentiating Equation 2.51 with respect to x and including the second boundary condition at the free end gives

$$0 = C \omega_n \left(\frac{\rho}{E} \right)^{1/2} \cos \left(\omega_n \left(\frac{\rho}{E} \right)^{1/2} L \right) \sin (\omega_n t + \varphi) \qquad (2.52)$$

For Equation 2.52 to be true, the argument of the spatial cosine term must be $\pi/2$, $3\pi/2$, etc. Consequently the jth mode longitudinal resonant frequency for a cantilever beam, in radians per second, is

$$\omega_n = \frac{(2j-1)}{2}\left(\frac{E}{\rho}\right)^{\frac{1}{2}}\frac{\pi}{L} \qquad (2.53)$$

Finding frequencies and mode shapes for other simple beam configurations requires only the use of their different boundary conditions. The overall approach can also be applied to other types of vibration. For example, similar arguments show that a beam under torsional oscillation about its longitudinal axis will be governed by the wave equation

$$\frac{\partial^2\theta}{\partial t^2} = \frac{G}{\rho}\frac{\partial^2\theta}{\partial x^2} \qquad (2.54)$$

2.3.2 Transverse vibration of beams

By a method broadly the same as that employed in the previous section, the governing equation for the transverse vibration of a beam can be derived from a consideration of the restoring forces if it is bent to a curvature R. A detailed proof using the method of Lagrange is beyond the scope of this book (see for example Goldstein, 1980, Morrison and Crossland, 1970, Timoshenko, 1955). It is simply stated that motion of the beam must satisfy the equation

$$EI\frac{\partial^4 y}{\partial x^4} = -\frac{M}{L}\frac{\partial^2 y}{\partial t^2} \qquad (2.55)$$

where M is its total mass and L its length. The general shape of this can be intuitively understood by comparison with Equations 2.15 and 2.48. Again, the solution form of Equation 2.49 can be assumed, whence we obtain, introducing a notational convenience,

$$\frac{d^4 f(x)}{dx^4} = \alpha^4 f(x) \quad \text{where} \quad \alpha^4 = \frac{M\,\omega_n^2}{LEI} \qquad (2.56)$$

The natural frequency is usefully expressed as

$$\omega_n = (\alpha L)^2\left(\frac{EI}{ML^3}\right)^{\frac{1}{2}} \qquad (2.57)$$

Values for αL will depend upon the boundary conditions for the particular beam and its supports. For example, with simple supports at each end there must be zero displacement and zero bending moment at each support. These conditions can be expressed mathematically from Equation 2.15

$$@ \ x = 0, \quad x = L; \quad f(x) = \frac{d^2 f(x)}{dx^2} = 0 \qquad (2.58)$$

A general solution to Equation 2.56 is given by

$$f(x) = C_1 \sin(\alpha x) + C_2 \cos(\alpha x) + C_3 \sinh(\alpha x) + C_4 \cosh(\alpha x) \qquad (2.59)$$

where C_1, C_2, C_3 and C_4 are arbitrary constants. Combining Equations 2.58 and 2.59 gives

$$@ \ x = 0$$

$$\frac{d^2 f(0)}{dx^2} = 0 = -C_2 + C_4$$

$$f(0) = 0 = C_2 + C_4$$

$$\therefore \ C_2 = C_4 = 0$$

$$@ \ x = L$$

$$\frac{d^2 f(L)}{dx^2} = -C_1 \sin(\alpha L) + C_3 \sinh(\alpha L) = 0$$

$$f(L) = C_1 \sin(\alpha L) + C_3 \sinh(\alpha L) = 0 \qquad (2.60)$$

For any finite length of beam the value of $\sinh(\alpha L)$ will be greater than zero, from which we must conclude that $C_3 = 0$. Additionally, for any finite value of C_1 (if this were also equal to zero, there would be no vibration) then $\sin(\alpha L)$ must be zero and so

$$\alpha L = j\pi \quad j = 1,2 \ldots \infty \qquad (2.61)$$

Combining Equations 2.49, and 2.59 with the boundary conditions for the simply supported beam gives its vertical displacement, y, at any position x and time t

$$y = C_1 \sin(\alpha x) \sin(\omega_n t + \varphi) \qquad (2.62)$$

where for successive vibrational modes, α is defined by Equation 2.61 and ω_n then found from Equation 2.57. The fundamental frequency is $\pi^2(EI/ML^3)^{0.5}$ and the next harmonic at four times that frequency

Different values for αL will occur with the different boundary conditions of other beam configurations. These are extensively tabulated, for example in Bishop and Johnson, 1960. A matrix solution is presented in Timoshenko, 1955.

2.3.3   Rayleigh's method

In the previous sections, it has been emphasized that free vibration occurs as a consequence of a periodic interchange of a constant mechanical energy between kinetic and potential forms, resulting in a sinusoidal variation in the displacement of a point in a structure. Fully analytic techniques can become very complex for 'real' structures and so simpler, approximate approaches are often sought. One of the most popular methods is that proposed by Lord Rayleigh, 1894. An initial guess of the mode shape of the vibrating element is made and used to estimate the potential (strain) energy. The natural frequency is then calculated by comparing the maximum kinetic energy, occurring at minimum potential energy, and the maximum potential energy, occurring at zero speed. Because energy in a closed system is always conserved, the magnitudes of the maximum changes in potential and kinetic energy must be equal.

Consider first of all a very simple example which has a well known solution. A mass, m, resting on a frictionless horizontal surface is attached to a rigid wall through a massless spring of stiffness λ. Its instantaneous position and velocity could reasonably be expected to follow simple harmonic motion described by the equations

$$x(t) \;=\; C\sin(\omega_n t)$$

$$\dot{x}(t) \;=\; C\omega_n\cos(\omega_n t)$$

(2.63)

The maximum kinetic energy of the mass occurs at the maximum speed, that is when the cosine term is equal to unity. The maximum potential energy is stored in the spring at instants when the mass is momentarily motionless, when the sine term is unity. So

$$T_{max} \;=\; \frac{1}{2}\,m\,\dot{x}_{max}^2 \;=\; \frac{1}{2}\,m(C\omega_n)^2$$

(2.64)

$$U_{max} \;=\; \frac{1}{2}\,\lambda\,x_{max}^2 \;=\; \frac{1}{2}\,\lambda C^2$$

(2.65)

Equating Equations 2.65 and 2.66 provides an expression for the natural frequency, in this case the familiar square root of the stiffness to mass ratio. This is an exact solution that demonstrates that the method works. Note, however, that as derived here it should be regarded as an estimate since it is based on the *assumption* that the mode shape of Equation 2.63 describes the motion when applying it to the calculation of T_{max} and U_{max}.

For transverse vibrations of continuous beams, the mathematics becomes a little more involved but the technique remains essentially similar. The first task is to choose the shape of the distorted beam. Fortunately, we already have a technique for deriving a plausible shape that will at least satisfy the boundary conditions for the beam; integrating Equation 2.15 for the bending moment of a beam under a uniform gravitational load. Consider a small element of mass δm as it is deflected y by gravity. As it displaces, the net force on it reduces uniformly from δmg to zero, so its potential energy will be $\delta mgy/2$. If the gravitational field were suddenly removed, we assume that the beam would oscillate about its undistorted position as a standing wave, retaining the same shape with an amplitude varying sinusoidally with time. If this is so, the maximum kinetic energy of the element under consideration would be $\delta m(\omega_n y)^2/2$. Integrating over the whole length of the beam, equating potential and kinetic energies and solving for ω_n we obtain

$$\omega_n^2 = g \frac{\int_0^L y\,dm}{\int_0^L y^2\,dm} = g \frac{\int_0^L y\,dx}{\int_0^L y^2\,dx} \qquad (2.66)$$

where the righthand form applies only to a uniform beam. Although g appears in this expression, it will cancel on substituting for the deflected shape y in the integrals.

For the uniform simply supported beam discussed in the previous section, the bending moment, BM^*, at any point x along the beam is given by

---

* In this section we use BM to denote bending moment to avoid confusion with the total beam mass, M. Unfortunately, there are bound to be occasional notational conflicts in a book of this type unless special, and unfamiliar, symbols are used. Our general policy has been to ease cross-referencing to other sources by using retaining commonly used symbols and relying on context to distinguish between, say time, temperature, torque and kinetic energy.

$$BM = EI\frac{d^2y}{dx^2} = \frac{Mg}{2}\left(\frac{x^2}{L} - x\right) \tag{2.67}$$

where M is its total mass and L its length. Intergrating twice and applying the boundary conditions that $y = 0$ at both $x = 0$ and at $x = L$, the gravitationally deflected shape is

$$y = \frac{Mg}{24EIL}(x^4 - 2Lx^3 + L^3x) \tag{2.68}$$

It is then readily found that

$$\int_0^L y\,dx = \frac{Mg}{24EIL}\cdot\frac{L^5}{5} \tag{2.69}$$

$$\int_0^L y^2\,dx = \frac{M^2g^2}{576E^2I^2L^2}\cdot\frac{31L^9}{630} \tag{2.70}$$

From which, on substitution into Equation 2.66,

$$\omega_n = 9.8767\left(\frac{EI}{ML^3}\right) \tag{2.71}$$

This is only about 0.1% from the exact value derived in Section 2.3.2, for which $(\alpha L)^2$ is 9.8696. Other values will be obtained for other beam configurations since they will have different boundary conditions and different bending moment distributions.

It is not absolutely necessary to choose the distorted shape to be that due to a static gravitational load. The strain energy stored in bending a prismatic beam can be derived generally from the integral

$$U = \int_0^L \frac{(BM)^2}{2EI}\,dx \tag{2.72}$$

and the kinetic energy from

$$T = \omega_n^2\int_0^L \frac{M}{L}y^2\,dx \tag{2.73}$$

40

It can be shown that the value for the first mode natural frequency estimated by Rayleigh's method will always be higher than that of the real system. The accuracy of the estimate will, of course, depend upon that of the original guess for the mode shape. Knowing this, iterative techniques can be used in which the intial choice is successively modified until the frequency estimates converge sufficiently to indicate that they are close to the exact value. These techniques are usually associated with the names of Rayleigh, Ritz, Stoddola and Vianello. However, in our experience, the deviation of the mathematical model from reality is often of far more importance than the slightly high dynamic response indicated by such solutions.

Rayleigh's method does not apply only to transverse vibration of beams, as was hinted by the initial illustration. (Incidentally, that may also be solved in terms of an assumed gravitational deflection of the spring.) It may be used wherever it is possible to provide reasonable guesses at the mode shape and associated potential and kinetic energy. Another classic application is to the longitudinal vibration of springs that have significant mass, see for example Morrison and Crossland, 1970.

References

Arya A.P., 1990, *Introduction to Classical Mechanics*, Allyn and Bacon, Boston.

Bishop R.E.D. and Johnson D.C., 1960, *Mechanics of Vibration*, Cambridge University Press, London, 355-407 (NB the symbol λ is used in this text instead of α)

Chorlton F., 1983, *Textbook of Dynamics*, Ellis Horwood, Chichester, UK.

Goldstein H., 1980, *Classical Mechanics*, 2nd ed., Addison Wesley, Massachusetts, chapters 1, 2 and 12.

Morrison J.L.M. and Crossland B., 1970, *An Introduction to the Mechanics of Machines*, Longman, London.

Pars L.A., 1965, *Treatise of Analytical Dynamics*, Heinemann, London, Chapter 8.

Lord Rayleigh, 1894, *The Theory of Sound*, MacMillan, London.

Ryder G.H., 1969, *Strength of Materials*, 3rd ed., Macmillan, London.

Timoshenko S. and Young D.H., 1968, *Elements of Strength of Materials*, 5th ed., Van Nostrand, Princeton.

Timoshenko S., 1955, *Vibration Problems in Engineering*, 3rd ed., Van Nostrand, Princeton.

3 FUNDAMENTAL CONCEPTS IN PRECISION DESIGN

This chapter introduces the basic concepts that underpin most precision mechanical design and that are much used throughout this book. An accurate device must be repeatable and stable. Central to this requirement, and given pride of place here, is the geometric notion of the kinematic constraint of ideal rigid bodies. It is applied to the determination of the relative motion of two elements, to the efficacy of positioning devices and to the behaviour of mechanisms. All real structures are elastic, which causes deviations from true kinematic behaviour, but which may also be exploited directly as pseudo-kinematic constraint. The various themes of kinematic design identify design approaches that are potentially precise, but they say nothing about the individual stability of single elements in a system. All controlled displacements depend on measurements relative to a datum, but forces that must be carried through a structure distort it and may move that datum. Hence the concepts of measurement and force loops are examined in some detail, as are the principles of alignment. Then some methods of improving loop performance are introduced. Nulling techniques may give highly precise control with few side-effects from non-linearity. Mechanical or computer-based compensation methods can reduce loop sensitivity to thermal disturbances and be used to separate predictable systematic errors. Finally, there are some observations on the potential benefits of symmetry and of scaling devices to smaller dimensions. Taken together, the topics covered provide a means of assessing proposals. A design that does not comply with the suggested practices is not necessarily wrong, but it is reasonable to require an explicit justification for the non-compliance.

Introduction

In the course of the history of precision instrument design, particularly over the last hundred or so years, many errors, blunders and faute de mieux solutions have been observed. There have also been many exceptional designs, often evolving from new insights into general design principles. A good knowledge of both ensures that present day designers are armed with a number of safety checks when assessing the feasibility of a proposed design. These are conveniently formulated as a series of fundamental questions such as:

1. Does it contain the correct number of kinematic constraints?

2. Where are the metrology and force loops and how are they shared?

3. Are there any heat sources and, if so, what are their likely effects?

4. Is the design symmetrical and are the alignment principles obeyed?

5. What other sources of error are likely to be encountered and can they be reduced by using compensation or nulling?

The exercise of finding, or failing to find, satisfactory answers to such questions involves a study of the proposal that adds greatly to our confidence that it is sound and filters out most poor suggestions. Experienced designers consider them as second nature and may not be aware that they are so doing. Do not worry if they seem perplexing at this stage. A major underlying theme of this book is concerned with their practical application. First, in this chapter we introduce and review the basic principles of precision design from which these and other questions arise.

3.1 Kinematics of constraint

Kinematics is the study of the geometry of motion, to be contrasted with kinetics, which studies the forces involved. It therefore deals with idealized concepts such as point contacts applied typically, for the purposes of this book, to the analysis of rigid body mechanisms. Kinematic constraint is concerned with the number of degrees of freedom possessed by a mechanism geometry. If the degree of constraint does not exactly match the required freedom, it is unlikely that the design will function as expected.

3.1.1 Coordinate systems

For most systems, known as 'holonomic', the number of degrees of freedom is equal to the number of coordinates necessary to fully define both the position and orientation of all its elements. We may use any coordinate system, for example Cartesian or cylindrical or spherical polar, but the number of coordinates remains always the same. The choice between representations may be made purely on the grounds of obtaining the simplest mathematical description.

Consider a very simple system such as the pendulum shown in Figure 3.1a. A mass, m, is attached to one end of a light, rigid rod. The mass is assumed to act as a particle located at its centre of gravity, G, so the rod has effective length L. The other end of the rod is attached by a simple, single axis hinge to a rigid support (shown shaded) so that it hangs in a vertical plane. Oscillations of the

a)   b)

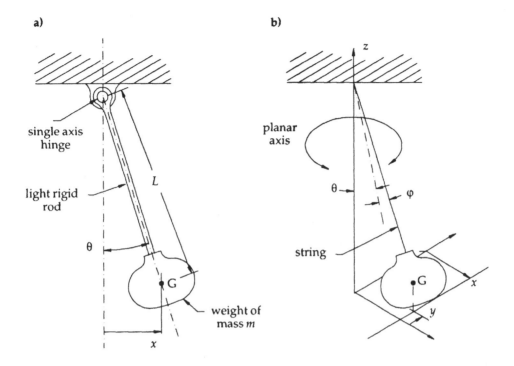

Figure 3.1   Simple pendulum a) A clock pendulum, b) Foucault's pendulum

pendulum in a gravity field are caused by the periodic exchange of kinetic and potential energies. Thus all that is needed to derive the equations of motion for the pendulum is an equation for the position of the centre of mass. There appear to be two convenient coordinates; the linear coordinate, x, and the circular coordinate, θ. The gravity field acts in (actually, defines) the vertical direction so the change in potential is related to the change in height, h, of the centre of gravity due to any displacement. The change in potential energy, U, relative to the lowest position of the mass can be expressed by either of

$$U \;=\; mgh \;=\; mgL[1 - \cos(\theta)] \tag{3.1a}$$

$$U \;=\; mgL\left[1 - \left(1 - \left(\frac{x}{L}\right)^2\right)^{\frac{1}{2}}\right] \tag{3.1b}$$

The above equations are identical and each involves only one coordinate, either x or θ. The simple pendulum is described as a single degree of freedom

45

mechanism. Computation of the velocity of the mass and thus its kinetic energy, T (= $mv^2/2$), can also employ either coordinate system. However, it rapidly becomes clear that the circular coordinate, θ, will provide considerably simpler solutions than the linear one, x, since the velocity, v, is given by the equations

$$v = L\,\dot{\theta} \tag{3.2a}$$

$$v = \frac{\dot{x}}{\left(1 - \left(\dfrac{x}{L}\right)^2\right)^{\frac{1}{2}}} \tag{3.2b}$$

If a second pendulum were suspended from the mass of this one, see Figure 2.5, two coordinates, most conveniently the angle of each rod from the vertical, would be needed. This two degree of freedom system illustrates the general rule that the freedom of a system is the sum of the freedoms of its elements.

In Figure 3.1b the same mass is suspended by a light string that is assumed not to stretch or contract, and is free to swing in any direction. Provided that the string remains in tension, it may be treated as a rigid rod suspended by a perfect ball joint. Either the coordinates θ and φ or x and y (or any independent combination of two of these) may be used to define the position of the mass, from which changes in potential and kinetic energy can be calculated and used with Lagrange's Equation, see Chapter 2, to derive its motion. Two *independent* coordinates are necessary to describe the motion of the mass, indicating that this pendulum is a two degree of freedom system. The motion can be qualitatively deduced by considering such a pendulum suspended above one of the Earth's poles. If it is freely swinging in one plane, there is no external force present to transfer linear momentum into other directions. Consequently, to an observer standing on the Earth, its periodic motion will appear to rotate about its axis once a day. The oscillation of the pendulum at other latitudes is more complex and beyond our present scope, but there is a strong intuition that spherical coordinates will provide simpler expressions than Cartesian ones. The study of its motion was used by Jean-Bernard Leon Foucault in 1851 to demonstrate the rotation of the Earth about its own axis.

Exceptions to the rule that the number of independent coordinates equals the number of freedoms usually involve rolling motion. As a simple illustration, consider the case of a circular hoop rolling on a plane, Figure 3.2. Generally, a full mathematical description of the hoop motion must explicitly include both its displacement along the inclined ramp and its rotation. It is essentially a single degree of freedom system, yet we may use a single coordinate, reflecting this, only in the specific instance that slip can be

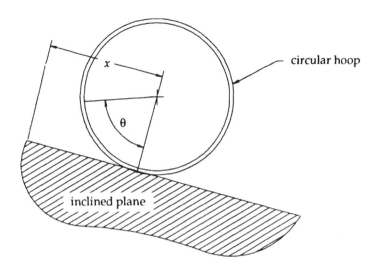

Figure 3.2 A non-holonomic system: a hoop rolling down an inclined plane

expressed directly as a function of displacement or rotation. For example, a no-slip system has the constraint that $\delta x = r\delta\theta$. Clearly, in such a simple case the constraint equation expressed in differential form can be integrated to give the explicit form, $x = r\theta$, from which the rotation of the hoop can be expressed in terms of its displacement, or vice versa. Generally, it will be possible to integrate any constraint equation involving only two coordinates. However, more realistically the hoop would be able both to roll in the xy plane and rotate in the other two axes. Under these conditions a more involved constraint equation including the six coordinates of its possible interelated freedoms is necessary and this cannot be integrated. Mechanisms involving non-integrable constraints which must therefore be treated as nonholonomic systems, do not occur in this book, so no further discussion is given here.

3.1.2 Basic point constraints

A simple single degree of freedom constraint may be envisaged as an infinitely rigid sphere pressed against an infinitely rigid flat, with a frictionless interface, Figure 3.3. The contact of real spheres is considered in Chapter 5, but this gross idealization is quite satisfactory for the kinematic modelling of instrument mechanisms and other machines where forces are low and hard materials can be used, see for example Whitehead, 1954, Slocum, 1988.

Assuming that there is no friction at the interface, the number of degrees of freedom can be deduced from the Cartesian coordinates shown in Figure 3.3.

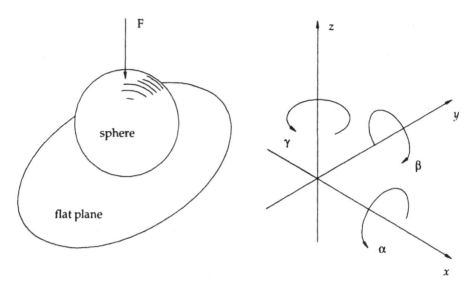

Figure 3.3 A single degree of freedom constraint provided by a sphere on a flat. The freedoms of the Cartesian coordinate system are also shown.

The sphere can rotate about any of the three axes and can slide in the x and y directions. Motion is restricted only in the vertical direction, z. In the absence of the plane the ball has six independent freedoms and introducing a point contact reduces that to five: it imposes a single constraint. This reasoning can be extended to give two very important lemmas*

1. *Any unconstrained rigid body has six degrees of freedom.*

2. *The number of contact points between any two perfectly rigid bodies is equal to the number of their mutual constraints.*

Models for predicting the motion of mechanisms can then be constructed in terms of the number of contact points, using the geometrical abstraction of spheres connected by rigid links. This is illustrated in Figure 3.4. Note carefully that in all cases a force in a suitable direction must be applied to maintain contact, as shown in Figure 3.3. This is often referred to as constraint with 'force closure'.

Using the notation of Figure 3.3, a two degree of freedom constraint can be constructed from two spheres joined together to make a rigid body and placed

---

* A lemma is an argument or proposition that forms the preliminary basis of a mathematical or literary discussion.

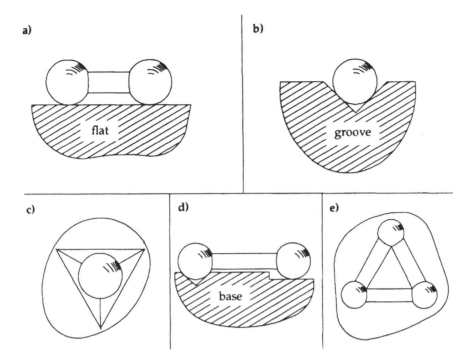

Figure 3.4 Idealised kinematic constraints; a) two balls rigidly connected and resting on a flat; b) a ball in a vee groove; c) a single ball in a trihedral hole, d) two balls rigid by attached with one resting in a groove and the other on a flat, e) three rigidly attached balls on a flat.

onto a surface, Figure 3.4a. The constraints are in the z axis and either the β or α axis. Alternatively, a single rigid sphere can be placed into a groove, Figure 3.4b, giving constraints in the z axis and either the x or y axis. Figure 3.4c to Figure 3.4e all represent three degree of freedom constraints. The constrained axes for these configurations are respectively; x, y and z; z, x or y and β or α; and z, α and β. Consideration of such configurations yields two more useful rules:

1. Each link will remove a rotary degree of freedom.

2. The number of linear constraints cannot be less than the maximum number of contacts on an individual sphere.

A trihedral hole, as shown in Figure 3.4c, is very difficult to produce using conventional manufacturing methods. For miniature structures one technique is to use an anisotropic etchant on a single crystalline material, Petersen, 1982. In larger engineering applications it is common to approximate a trihedral hole

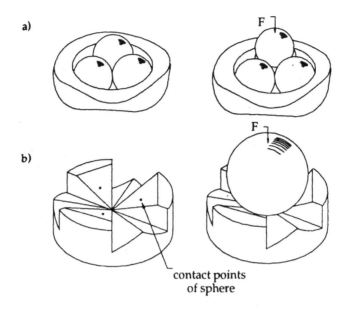

contact points
of sphere

Figure 3.5   Methods for achieving trihedral contacts

by other structures. Two methods are shown in Figure 3.5. Figure 3.5a consists of three spheres pressed into a flat bottomed hole. A further sphere placed on these will contact each at their point of common tangent. The triangular hole of Figure 3.5b can be produced by machining three coplanar grooves at a pitch of 120 degrees using a vee shaped cutter that is angled so that one of the faces is vertical and the other is inclined at the full angle of the groove relative to the workpiece surface. The three contact faces will slope out from the common centre of the grooves on the inclined planes.

Another apparently simple method for locating a sphere is to use a drilled conical hole. Theoretically this provides a line contact. In reality location is poor because neither a sphere nor a conical hole can be produced without finite surface finish and errors of form. A common form error with drilling makes the conical hole slightly oblong. As the sphere approaches the hole, it makes a first point contact and then slides down the cone until another opposing contact provides an equilibriating force. The joint can then support a vertical load, but there is only a two point contact, leaving a linear degree of freedom along the major axis of the elongated circular section. The sphere is free to rock either way until it makes a third contact, so at least two sets of kinematic location points will be present. Thus a sphere and cone mount tends to be unstable unless sufficient load has been applied that the distortion of the contacts is greater than the imperfections of the two surfaces.

CONSTRAINTS	CONFIGURATION
1	Ball on flat
2	Link on flat Ball in groove
3	Ball in trihedral hole Link with one ball in groove and other on flat Three linked balls suitably distributed on a flat
4	Link with one ball in trihedral hole and the other on a flat Link with two balls both in a vee groove Link of three balls with two on a flat and one in a groove Link of four balls suitably distributed on two inclined flats
5	Link of two balls with one in a trihedral hole and the other in a vee groove Link of three balls with two in vee grooves and one on a flat Link of four balls with one in a vee groove and three on a flat Link of five balls suitably distributed on two inclined flats

Table 3.1: Idealized constraint geometries for the contact of rigidly connected spheres on a combination of flats, grooves and trihedral holes

3.1.3 Point constraints and relative freedom

Using the rules for linking balls and counting contacts, many geometries that constitute kinematic constraint can be invented to provide requisite freedom of motion between parts of a mechanism. Calling a pair of rigidly connected balls a link, some combinations giving from one to five degrees of constraint are listed in Table 3.1. For certain configurations it is difficult to envisage the geometry associated with the contact conditions specified. This is because there are geometric singularities in the rules and simply counting the total contacts is not adequate. Consider the last configuration mentioned in Table 3.1. This could be read as five balls on a flat surface. However, it is clear that such a geometry allows sliding on the plane and so has a least two degrees of freedom, so there is an inconsistency. The three-legged 'milking stool' demonstrates that three contacts provide stability of a rigid body on a surface. Any more legs on a stool will result in *overconstraint*. In general this will lead to an instability through rocking about two legs to switch between different sets of three contacts. Four or more legs can only be in contact simultaneously if either they happen to have exactly the right lengths or there is (elastic) distortion in the them. The former is pure luck, given manufacturing uncertainties, and the latter may be undesirable in precision systems where we wish to keep stress levels as low as possible.

We extend the earlier rules. A specified relative freedom between two subsystems cannot be maintained if there is underconstraint. Overconstraint does not prevent the motion but should be avoided unless there are special reasons for it. (A typical case might be where a chair uses four or more legs to provide more stability against off-centre loads and either slight rocking is tolerated or adjustable legs incorporated.) Underconstraint means that there are too few *independent* contacts.

A system might contain exactly the number of contacts corresponding to the number of freedoms to be constrained, but if there is local overconstraint those contacts are not independent and the overall system is underconstrained. Getting these conditions correct requires design skill, not mere rule following.

A single plane can support at most three constraints, for it provides no way to prevent either translation or rotation parallel to it. To implement a single degree of freedom system with the five single point contacts it is necessary to distribute them on two non-parallel flat surfaces, for example as shown in Figure 3.6. The three linked spheres contacting one flat surface are rigidly

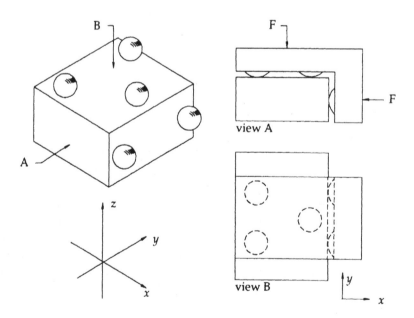

Figure 3.6 A single degree of freedom movement consisting of five point contacts on a straight prismatic vee slide

attached to two more spheres which contact a second, perpendicular face. Typically, as shown, all the spheres would be embedded into a solid carriage. Its degree of freedom can be deduced by considering the loading necessary to keep all five spheres in contact. Firstly, the three point support could be positioned onto the horizontal plane resulting in a linear constraint in the z axis and rotary constraints about the x and y axes. A carriage placed on this plane is free to slide in the x direction until either of the two remaining spheres contacts the vertical face. The x axis linear freedom is then constrained. Further horizontal force would cause the carriage to rotate until the fifth sphere comes into contact, removing the rotary freedom about the z axis. Consequently the mechanism gives a correctly constrained single degree of freedom linear motion along the y axis. As just analysed, two forces were introduced to provide the closure. Obviously they may be added vectorially to act at the corner, and if the whole mechanism is rotated in the plane A, gravity can provide all that is needed. The mechanism is a special case of a groove or prismatic vee and in practice an almost arbitrary shape may be chosen provided that an angle remains between the planes. Too small an angle will result in very high normal forces to provide adequate components to maintain the location. Note also that because there must be friction at real contacts, which can interact with other imposed forces to cause torques that tend to lift the contacts, it is preferable to employ body forces, such as weight, rather than tractions for force closure.

So far we have considered only the contact of two rigid bodies through spheres firmly attached to them. Kinematic constraint can also be achieved through spheres that remain free floating between the two bodies. The classic screw and diameter measuring instrument shown in Figure 3.7 is a typical example of an instrument utilising free rolling spheres. A bar or screw thread to be measured is clamped between two centres on a rigid base. A micrometer and fudicial gauge are mounted on a moving stage having a single linear degree of freedom perpendicular to the axis of the clamping centres. The fudicial gauge is used to ensure that the correct force is exerted by the micrometer spindle onto the bar being measured. Two parallel grooves have been precisely ground in the base and a further groove machined into the underside of the micrometer carriage.

Two spheres are placed in the left hand vee groove, towards either end, and a third is placed centrally in the right. There are a total of five contact points between the carriage and spheres in a geometry that correctly leaves a single degree of freedom. Consequently, the three spheres could be rigidly connected to the base and the carriage would slide over them in a stable motion. However, in practice, frictional effects at the sliding interfaces will impose an undesirable additional force on the bar from the micrometer spindle. This force is reduced by allowing the spheres to roll, see Chapter 5, by resting them in the grooves of

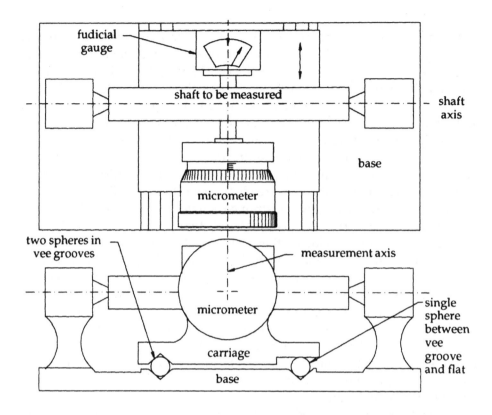

Figure 3.7 A schematic diagram representing two orthogonal views of a screw and diameter measuring instrument

the base. As there are six contact points on the base there must be some sliding unless the vee grooves are perfectly parallel, but the total frictional force will be much smaller than with pure sliding. Grooves are needed to keep the balls in place and, to prevent them from migrating along the grooves, pins are inserted to allow only a few millimetres of free rolling. This range is quite adequate to provide improved performance.

3.1.4 Kinematic clamps

So far, there has been no discussion of mechanisms involving constraint in all six degree of freedom between two rigid bodies. They are unable to move relative to each other and are considered 'clamped'. An ideal kinematic clamp has six independent contact points. Any more or less results in over- or under-constraint whereby there will be redundancy or a deficit in the design. The concept of the kinematic clamp is important for two reasons. Since in mounting the parts the contacts will be made sequentially (see the discussion in Section 3.1.3) an exactly constrained pair will always locate together stably without strain even in the presence of poor manufacturing tolerances. Only gross dimensional errors will prevent this. The relative alignment will vary slightly with the tolerance, but usually the benefits of a firm, low stress and relatively low cost clamp outweigh this disadvantage. Also, many cases exist where a workpiece or instrument must be removed from its original location and subsequently relocated as near as possible to its original position. A familiar example is that of the surveyor's theodolite, but there are many others. Here the very reasoning that the kinematic clamp finds its own position implies that it will find the same position every time, provided it has not been distorted.

There are many geometries that provide the necessary six contacts, but the types I and II Kelvin clamp shown in Figure 3.8 are by far the most common. Both clamps consist of three rigidly connected spheres mating against a trihedral hole, groove and flat in the type I clamp and against three vee grooves at 120° pitch angle in the type II clamp. The type I geometry offers a well defined

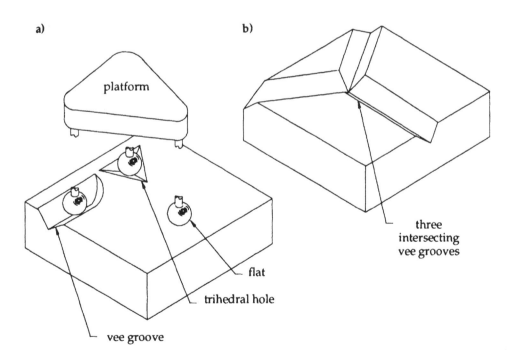

Figure 3.8   The kinematic clamp a) Type I (the connections between the spheres and platform have been sectioned to reveal the contact points), b) type II

translational location based on the position of the trihedral hole, but is asymmetric. The type II geometry is easier to make and has the distinct advantage of a centre of symmetry. Any expansion of the base plate or platform will not, within reason, affect the position in the horizontal plane of the centroid of the three spheres relative to the centroid of the grooves, so the location is not very susceptible to thermal variation. For this reason it is a popular choice for applications such as theodolite mounts and other instruments that operate in a varying environment. Note that the symmetrical groove pattern confers its own advantages but is not a kinematic requirement: any set of grooves will do provided that they are not all parallel.

3.2 Mobility and the kinematics of mechanisms

Just as the idea of kinematic constraint of a body is a useful idealization, geometric abstractions may be used to analyse kinematic chains, that is sets of rigid bodies connected by ideal joints of various types. A mechanism is a kinematic chain in which the relative positions and orientations of its members provide useful information and, as such, one of its members must be rigidly attached to our frame of reference. Even a purely geometrical description of interlinked members rapidly becomes difficult as the mechanism complexity rises, so there is a tendency for formal texts on mechanism design to be treated as applied mathematics, while practicing engineers adopt empirical methods. We shall concentrate only on the analysis of freedoms of proposed designs, reviewing briefly the idea of mobility of mechanisms first in general and then, with a few extra details, in the special case of planar mechanisms. A rather quirky but engineering-based examination of the subtleties of this subject is provided by Phillips, 1984.

3.2.1 Mechanism Mobility

Since a rigid body has six degrees of freedom in space, a set of n bodies has $6n$ freedoms. If the bodies are elements of a kinematic chain, the connections between them reduce the total number of freedoms, or independent coordinates needed to fully describe the position and orientation of the set. For example, if two rods are connected by an ideal ball-joint then we need six coordinates to fix one of them but, since the end position of the other is now defined, only three angular coordinates to fix the other: the combination has nine freedoms overall. Each joint can be considered as either providing f freedoms compared to a totally rigid connection, or c constraints compared to a separation. An ideal simple hinge provides one rotational freedom (5 constraints) and the ideal ball-joint three freedoms (3 constraints) Always

$f = (6 - c_i)$. Additionally, by our definition of a mechanism, one of the members is fixed and so has no freedoms.

We now define the mobility, M, of a mechanism to be the total number of freedoms provided by its members and joints. This equates to the number of independent controls that must be applied to the mechanism to fully define the position of all its members. In general a mechanism will have n members, of which $(n - 1)$ provide freedoms, and j joints each of which provides c_i, or $(6 - f_i)$, constraints. This leads to the expression, sometimes known as the Kutzbach criterion,

$$M \; = \; 6(n - 1) - \sum_{i=1}^{j} c_i \tag{3.3}$$

or, alternatively,

$$M \; = \; 6(n - 1) - \sum_{i=1}^{j} (6 - f_i)$$
$$= \; 6(n - j - 1) + \sum f \tag{3.4}$$

where $\sum f$ indicates the total freedom of all the joints.

A negative mobility indicates kinematic overconstraint and so provides a warning that all is not well in the mechanism design (but see Section 4.3.3). We hope that mobilities of zero, one or two would confirm a design intention to produce a statically determinate structure, a single controlled motion or a differential mechanism, respectively. Unfortunately, this is not always the case. If the mobility is not the correct value for the required motion, there is something wrong, but we may not conclude that the design is good just because the mobility has the expected value. Examples of why this is so will be deferred until simpler geometries have been discussed in the next section. Further, it may not be clear what constraints are actually active in some types of systems. Consider, for example, using a rod to prop open a trapdoor hinged to a horizontal floor, a situation that should be statically determinate. To avoid possible slipping of the ends of the rod, a kinematic design might provide it with ball ends and locate these in trihedral holes in the door and the floor. Intuitively such an approach feels fine. However the system has three members (remember the floor!) and three joints. The hinge has one freedom and each end of the rod three, so Equation 3.4 indicates a mobility of one. A closer examination of the system resolves the apparent paradox: the rod is free to rotate about its own axis, a motion probably irrelevant to our purpose. The system is indeed a single degree of freedom mechanism, but it is debatable whether extra expense should be incurred to include a keyed shaft for this application.

We do not advocate the use of mobility as a means of attempting to confirm a design. Rather we suggest that in calculating the mobility of a design and then reconciling the result with the behaviour expected through physical intuition, a better understanding of the system will emerge. It is this understanding that should lead to improved designs.

3.2.2   Kinematics of plane mechanisms

Very many real mechanisms are required to provide a motion that is, ideally, constrained to lie in a single plane. It is then often possible, and through the simplification it provides very convenient, to use a planar design. This makes use of the kinematic idealization that only the cross-section of the mechanism in the plane of its required motion need be considered. The simplification arises because, with elements confined to a plane, each has just three freedoms, two translational and one rotation. Single freedom joints involve either a rotation or a translation in the plane, and there could be a two-freedom joint combining these two types. The simpler design and analysis is obtained at the expense of additional deviations between the ideal and real situations. For example, rotary joints carry an implicit assumption that they are perfect hinges with axes always perfectly perpendicular to the design plane. In practice we often get sufficiently close to the idealizations that we may consider the out-of-plane motions as parasitic errors from the ideal rather than truly defined motions in three dimensions.

The number of coordinates required to describe the planar motion of a mechanism is, again, its mobility, M, now defined from Grubler's Equation

$$M = 3(n - 1) - 2j_1 \qquad (3.5)$$

where n is its total number of links, including the base, and j_1 the number of joints, each allowed just one degree of freedom. This form is given because two-freedom planar joints rarely occur in our context. The derivation should be clear by comparison with Equation 3.3.

It must be stressed that, as in the general case, there are a number of geometries for which this analysis does not work and there is no systematic method for determining the validity of a given result. Erroneous results often occur with structures containing a large number of parallel members and equispaced hinges. The most common cause of false results is for there to be sub-mechanisms that are respectively overconstrained and underconstrained such that the overall mobility appears to be correct. Checking individual loops helps to avoid this. The mechanism shown in Figure 3.9 has 5 links and 6 single

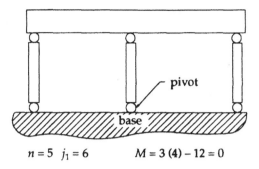

$$n = 5 \quad j_1 = 6 \qquad M = 3\,(4) - 12 = 0$$

Figure 3.9 An overconstrained linear mechanism

degree of freedom joints, giving a mobility of zero. However, providing the joints have some compliance, it would be relatively easy to use as a parallelogram motion. Physical intuition would allocate a single freedom to the device. This conflicts with the calculated mobility and warns of a potential design fault. Correct operation requires either give in the joints, leading to internal stresses, or that the dimensions of the legs are identical.

Figure 3.10a illustrates another important case. It appears to have 6 links and 6 joints, indicating a mobility of three, which is clearly incorrect. Care must be taken where three or more links meet at one point. It is not valid to attach more than two links to a joint and a more precise representation is shown in Figure 3.10b. There are now 8 joints and 6 links giving a mobility of –1 which reflects that fact that this is a solid structure with one unnecessary link. Further examples of analysis and the deliberate use of overconstraint are included in Section 4.3.3.

a) incorrect   b) correct representation

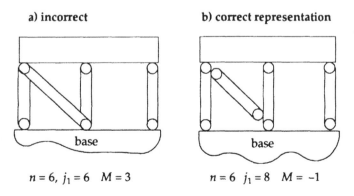

$$n = 6, \; j_1 = 6 \quad M = 3 \qquad\qquad n = 6 \quad j_1 = 8 \quad M = -1$$

Figure 3.10 The correct representation of a mechanism in which more than one link attaches to a hinge

3.3   Pseudo-kinematic design or elastic design

3.3.1   Load bearing, area contact and elastic averaging

The idealized point contacts of kinematic analysis would lead to infinite contact stress in the presence of a finite force closure. It is intuitively obvious that even with real ball contacts the stresses under loading will be very large and cause significant elastic and, ultimately, plastic distortion of the contact region. Large loads necessitate the use of large areas of load bearing surfaces. Chapter 5 investigates the nature of some real contacts, but a few points are worth raising here. As two rough surfaces (all real surfaces are rough) are brought together they first contact at the most prominent conjunction of their asperities. As load is applied, the peaks squash, and probably slightly less prominent features also touch and start to squash, until there is sufficient area to support the load at the maximum stress that the materials can carry. The real area of contact is generally a small fraction of the apparent planform of the contact area.

Consider the effects of this asperity distortion on the friction of the contact. Under a load, P (say), the real area of contact is proportional to the hardness of the softer material. In sliding, the frictional drag force, F, required to move one surface over another is that needed to shear the contacts, that is simply the product of the minimum shear strength, s, and the area of true contact, A. The coefficient of friction, μ, is defined as the ratio of the horizontal drag force to the normal load. For many metals, the hardness is approximately three times the yield stress, Y, and the shear strength about half the yield stress. So if the contacts are clean and devoid of contaminates we might expect

$$\mu \; = \; \frac{F}{P} \; = \; \frac{As}{AH} \; = \; \frac{Y}{2} \cdot \frac{1}{3Y} \; = \; \frac{1}{6} \tag{3.6}$$

This implies that the coefficient of friction is independent of load and geometry, which is another way of stating the famous Amonton's laws of (Coulomb) friction, and largely independent of material amongst metals. In reality, the many factors ignored in this crude model, such as contaminates, rate related effects, thermal, mechanical and chemical properties of surfaces, which cause variations in actual values and make them very difficult to predict. Typically coefficients of friction are in the range 0.1 to 1 for many materials. The model is sufficient to indicate that since friction tends to limit the repeatability of kinematic locations, the use of hard surfaces with low interface shear strength is ideal. This is achieved by introducing a surface film of fatty acids within oils (lubrication) or by coating the contacts with thin polymeric or low strength metallic or lamellar films.

Another implication of the small real contact of rough surfaces is that joints may have very low stiffness compared to the substrate materials. It may be necessary to apply a large preload to distort the two surfaces until there is a sufficient number of contacts to create a stiff interface between the two conforming bodies. In many other systems heavy external loads must be carried by mechanisms and so large nominal contact areas must be provided. Theory provides little help since contact between two elastic surfaces is complicated by both the surface finish from processing and overall shape of the surfaces. These factors and the effects of plastic distortion mean that only simple models of surface contact are mathematically tractable, Archard, 1957. Yet again the designer is left to his or her own intuition. In general a generous area is provided, but there is then the problem of ensuring that, say, a slideway and carriage conform without excessive drive forces being needed. This leads to the most common, if empirical, expression of the principle of semi-kinematic design: treat the centroid (since the actual centre of real contact is unpredictable) of each separate area of contact (or contact pad) as the position of a single ideal point contact and apply the rules of kinematic design as normal.

In general, large forces are not a desirable feature of ultra-precision designs. Loads must be supported and clamps maintained, but care should be taken, often through kinematic design, to minimize the generation of secondary forces. Large forces can be locked up in a rigidly clamped structure and increase the magnitude of such undesirable instabilities as creep and the effects of thermal mismatch between different materials. In situations such as vacuum flanges and other types of seal, a direct compressive force must be applied which could cause excessive bending moments. This can be largely avoided by incrementally tightening the fasteners during clamping. Consider the circular flange shown in cross-section in Figure 3.11. A rather poorly machined face (drawn with the manufacturing errors exaggerated) is to be bolted down to a flat one. After the first bolt is tightened the structure will be tilted as shown. If the second bolt is then tightened until the other side of the flanges touch, there will clearly be an asymmetric loading force around the circle of contact. Perhaps more importantly, the clamping force, F, of the second bolt will impose a bending moment of magnitude FL. This moment will cause a stress in the flange that will be concentrated in the fillet, at the position indicated, resulting in enhanced stress induced corrosion, reduced fatigue resistance and so on. Also the second bolt may not be able to provide sufficient force to overcome the bending stiffness of the flange to allow proper contact at the left hand side. There must be sufficient contact area at all positions around the flange to give a good seal. Both problems are reduced by securing the flange incrementally as shown in Figure 3.11b. Here one side of the flange is clamped into light contact but no further tightening of that bolt applied until the second bolt has pulled

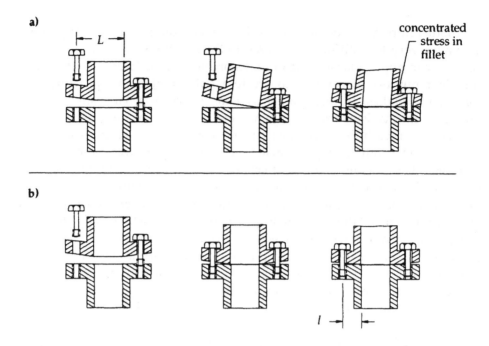

Figure 3.11 Fixturing of flanges and seals; a) Incorrect series clamping, b) Correct incremental clamping

the left hand side into light contact. The bolts are then alternately tightened a little until the required clamping force, F, is attained at each fixturing point. Using this method, the remaining bending moments at the clamp points are at maximum only Fl and evenly distributed about the contact circle. The illustration used only two bolts but more would be used in practice. The key point is that design and assembly must be considered together if best performance is to be obtained.

The discussion so far assumes that true kinematic constraint is always a desirable target and that the number of independent contact points between bodies required to support any applied loads is so defined. There are, however, instances in which we wish a load onto a rigid body to be distributed over a larger number of points. A rather simple example is the four-legged chair. Each leg can be considered as single contact on a flat surface, the floor. As only three constraints can be defined in a plane, Section 3.1.3, one leg is redundant. The overconstraint is deliberate because balance under eccentric loading is considered more important than highly precise location, a slight rocking may be tolerated. In reality, the overconstraint is often removed by placing the chair on rough ground and moving it about until all four legs touch or by pushing a wedge (or book) under the leg that is too short, both of which effectively supply an out of plane contact. Alternatively it may be placed on a compliant surface such as a carpet or, more generally, by designing the chair with compliant legs that will bend enough for all four to make contact. This approach which relies on the elasticity of the structure is used in many systems and is known as *elastic design* or *elastic averaging*. Note that a different use of elasticity in design is discussed in Chapter 4. The contact of real solids relies on small scale plastic distortion of materials. This is an inevitability rather than a design strategy, but sometimes yielding can be directly exploited to give *plastic design*. All real structures involve elastic and plastic effects and the terms should be reserved for occasions when deliberate departures from kinematic principles are involved.

3.3.2   Kinematic analogy to real systems

To further illustrate these concepts, consider the design of a rotary shaft. Kinematically, one such solution is to have a rigid link between two spheres, one constrained in a trihedral hole and the other in a groove, Figure 3.12a. The three contacts with the hole remove the three linear degrees of freedom and the two further contacts on the groove remove two rotary ones. The five contact system is a correctly constrained single axis rotation. One might think of using two trihedral holes to locate the two spheres. This geometry is over constrained and would rotate but only with poor repeatability. The shaft can be seated properly in the holes only if the length of the separating link is *exactly* correct. It might be forced in by bending the shaft at the cost of high bearing forces and poor balance. Engineers cannot work to exact dimensions and close tolerancing increases costs so this is not economically feasible. This emphasizes a general rule that

> *as the rules of kinematic constraint are compromised, manufacturing tolerances must become more exacting if mechanisms are to function in a satisfactory manner.*

or, alternatively,

> *a divergence from pure kinematic design results in increased manufacturing costs.*

A design decision that is all but trivial in a purely kinematic analysis, can be surprisingly easily lost in the complexities of real systems. If the shaft is to support a load, then the point contacts become impracticable and the ideal locations must be replaced by bearings of some sort. Ideally, following the principles of kinematic constraint, a bearing having three degrees of freedom should be used at one end of the shaft and one with four at the other end. It may

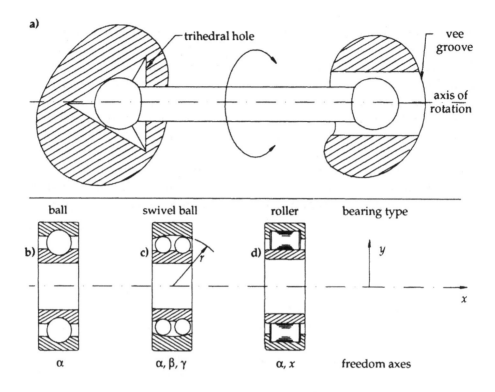

Figure 3.12 Rotary motion; a) Kinematic representation of a system having five constraints, b) A variety of rolling element bearings and their respective freedom axes

prove impossible to create such a design with sufficient rigidity for a specific application and compromises must be made. Figure 3.12b-d shows a range of typical rolling element bearings and their respective axes of rotation or translation. The simple ball bearing is effectively a five constraint mounting. The inner race of a swivel bearing can rotate about all three axes relative to the outer race so it can be considered kinematically equivalent to a sphere in a trihedral hole. The inner race of the roller bearing is free to rotate about and translate along the x axis and is often regarded as equivalent to the sphere in a straight vee groove. Actually it is quite strongly constrained in the other rotary freedoms, as the ball and groove is not, and so only functions as an equivalent if its inner race axis is well aligned to the predefined rotation axis. Our first thought might have been to use two ball bearings to support the shaft, which is seen to be highly overconstrained. Then a swivel ball and roller combination may be chosen, Figure 3.13a, but this still has seven constraints and one

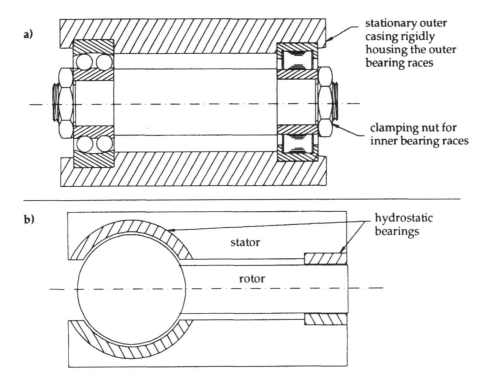

Figure 3.13 simple rotary shaft: a) Using rolling element bearings, b) Utilising journal or hydrostatic bearings.

freedom. We have to rely on close tolerance alignment with a tiny amount of elastic movement taking out the residual errors. For this reason the bearing industry has regularly been one of the leaders in precision manufacturing.

As loads increase still further, rolling element bearings become impractical and it necessary to resort to plain hydrodynamic or hydrostatic bearings (these are further discussed in Chapter 9). Again simple geometrical shapes are used to obtain a solution not too far from the kinematic one, but some degree of overconstraint is almost always inevitable. Figure 3.13b shows a typical example where, provided the pads are fairly short compared to the shaft, the internal clearances should allow enough self-alignment to obtain a close approximation to the kinematic design.

It is common practice to use paired ball bearings because they are lower cost items than other rolling element types. This may be justifiable although it requires the axes of the two bearings to be coincident since even the arrangement in Figure 3.13a requires that the bearing centres lay on the defined axis, so both approaches incur significant manufacturing costs. Designs using rolling element bearings tend to be ten to a hundred times more expensive than a simple kinematic solution. For rather different reasons, those with hydrostatic and hydrodynamic bearings cost roughly the same amount again.

3.4   Measurement and force loops

By applying Newton's third law successively at different sections of a static structure that carries a load, we rapidly discover that to maintain equilibrium balancing forces can be traced around a closed path, perhaps involving the ground, to the point at which we started. Even in machines that have relative motions it is rare to wish the total structure to accelerate across the room and so there must again be a balance of forces as seen by an exterior observer. Closed paths, or loops, to carry the forces must be present. Giving a little thought to the nature of real measurements, we reasonably conclude that *all meaningful measurements are made relative to a datum or control*. Again there must be a continuous path, in some sense, to link the measurand and the datum. In mechanical measurements this path is likely to be a loop contained largely within the machine structure.

The simple idea of loop thus emerges as a powerful concept for analysing design proposals. By definition, any force loop is stressed and therefore strained. The dimensional stability of a system under a given range of loading might be improved if it can be reorganized to make the principal force loops smaller in size. Measurement, or metrology, loops are critical to the degree of positional control that can be achieved. They must be identified and designed for stability. This follows from an important general rule:

> *any changes that occur to components in a measurement loop will result in changes in measured results that are not distinguishable from the measurement.*

Consequently, all errors occur in the loop of the system and its identification must always be the first stage of an error analysis. The key problem in all outline designs is to identify datums and the routes by which they are referenced.

The concept of loop is quite abstract, yet its application to design is very pragmatic. Thus we choose to explore the implications of both metrology and force loops for precision design through an extended example. Consider a simple (fictitious) stylus instrument for the measurement of surfaces, Figure 3.14. This consists of a stylus attached to the end of an arm that pivots about a knife edge. On the other end of the arm a piece of ferrite is positioned inside a coil to form an inductive displacement sensor. Displacement of the stylus

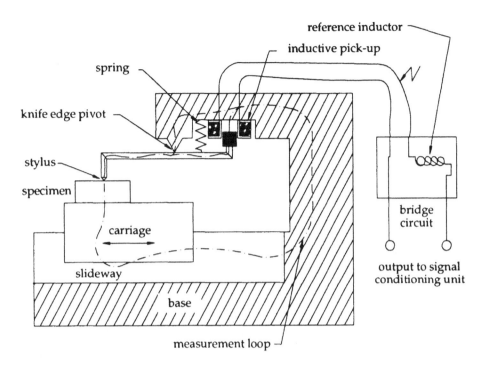

Figure 3.14 Schematic diagram of a simple surface measuring instrument

causes changes in the inductance of the coil. Changes in the inductance are detected by a simple AC bridge circuit. A specimen to be measured is placed on the carriage and the stylus brought into contact (the adjustment and other details have been omitted for clarity but would have to be considered in any real design). The carriage is then traversed in a straight line along the slideway and the variations in the output from the inductive gauge taken to represent the profile of the surface finish of the specimen.

The main mechanical measurement loop is quite easy to identify, Figure 3.14. It passes from the specimen to the stylus and sensor, then through the instrument frame to the carriage and so back to the specimen. Notice a small subsidiary path from the stylus arm through the pivot to rejoin the main loop beyond the coil. Its presence reminds us that the relative positions of the coil and pivot form the datum for determining the lever movement.

Having identified the measurement loop, we attempt to visualize how the behaviour of its components may influence the output. Initially, assume that the stylus contacts a specimen that is perfectly flat, infinitely hard, has zero thermal expansivity and that the instrument carriage remains stationary. The instrument will be situated in a laboratory or factory where the environment will change in both temperature and humidity and there will be vibrations due to surrounding machinery. Changes in the temperature will cause dimensional changes of all components in the loop. Unless they all cancel out around the loop, which is unlikely in any situation involving dynamic temperature variation, its overall dimension will change, which the output will reflect as a movement of the stylus. The components will have different thermal capacities so the rate of change of temperature will result in additional variations in the output. Most materials experience a change in elastic modulus with temperature (thermoelasticity) as well as a reduction in hardness. Two elements in this design may be affected. A change in hardness of the knife edge will alter the elastic or plastic distortion at the contact interface. Not only might this be monitored by the transducer as it moves the arm but it could have a 'knock-on' effect in that changes at the interface will change the extension of the spring and consequently the stress around the force loop (to be discussed shortly). The thermoelastic coefficient of the spring, as well as its thermal expansion, can introduce further errors.

If the stylus contact force varies the overall loop will deflect and cause a change in output. Loop compliance is likely to be dominated by bending in the stylus arm and the carriage bearings. Other effects include sensitivity to vibrations at frequencies near to resonances of the loop or sub-loops and interference in the wires leading from the instrument to the signal conditioning electronics. A full evaluation of potential error sources would require an unduly lengthy discussion and there is sufficient here for illustration. Of course, some, perhaps most, of them might be negligible, but we cannot know until we have thought of them and checked. However, recalling that all meaningful measurements are relative to a datum, we must not overlook the electrical measurement. Its datum is the reference inductor. A simple wire-wound inductor will have a thermal coefficient, probably of the order $0.1\% \ K^{-1}$, which limits the ultimate stability of the overall instrument. All other errors within the loop increase the uncertainty of the measured value, but only those of a similar (or larger) magnitude will do so significantly.

Now we must envisage the effect on the measurement if the instrument is allowed to scan a real specimen. The slideway will normally consist of a rubbing or rolling bearing having a characteristic noise resulting in non-rectilinear motion which will also be transmitted to the stylus. The stylus is small, so even at low loads the contact pressures can be high and the frictional and dynamic interaction between the stylus and specimen may be important. This will result in forces around the loops which are related to the surface being measured, Chetwynd *et al.*, 1992, Liu *et al.*, 1992.

With so many possible error sources one is tempted to ask whether the concept of measurement loops can increase our ability to assess an instrument at all. The answer to this question is probably best illustrated by considering an alternative design shown in Figure 3.15.

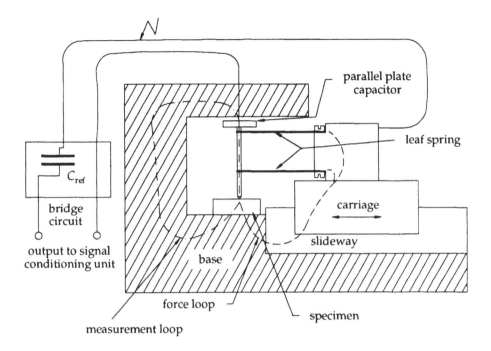

Figure 3.15 An alternative design for the simple surface measuring instrument of Figure 3.14

In this particular design the stylus assembly has been mounted onto the carriage and the stylus held by two leaf springs which are free to move in a parallelogram motion, see Chapter 4. The transducer is a parallel plate capacitive micrometer with one electrode attached directly onto the stylus shank and the other being a datum surface intrinsic to the instrument body. Assuming the stylus support springs are sufficiently compliant, contact would be maintained by the gravitational force. Similar arguments with respect to thermal, mechanical and electronic properties as possible error sources apply to this design as to the other. However, in this case, the measurement loop is shorter and may carry lower internal forces, so its dimensional stability should be better and it may have higher resonances that are less likely to pick up significant vibration amplitudes. The slideway errors no longer appear in the loop. This is an example of *decoupling* whereby possible error sources are removed from the measurement loop. Unless the stylus support springs have reasonably low stiffness, there could be a second order coupling of the motion error through a change of force in the measurement loop. Sensors are examined in Chapter 7 and here we merely note that an inductive gauge could still be used in this configuration. A motivation for using capacitive micrometry is that extremely good temporal stability reference capacitors can be produced with a temperature coefficient of better than 10^{-7} K^{-1}. Ultimately the choice between such rival designs would rest upon compromises between the overall costs, the level of performance required in practice, convenience of use and numerous other pragmatic considerations.

So far little mention has been made of forces in the structure. Any force in a measurement loop will cause it to distort and so manifest itself as a parasitic error. Note, incidentally, that in many loops deflection from bending is more significant than changes in axial dimensions of loop components. In the designs of Figure 3.14 and Figure 3.15 the presence of a spring implies forces within the structure in addition to gravitational effects which we hope will remain constant. The slideway drive, although not shown explicitly, will also involve forces. Consequently, all the elements through which varying forces are transmitted should be identified and the force loops mapped as part of the analysis. Figure 3.15 indicates the main force loop in that instrument. The forces are caused by distortion of the stylus support spring and these are transmitted through the carriage and slideway and into the specimen via the base. Forces from the carriage drive just pass through the bearings to the base and back to the actuator. A primary consideration in any design is deciding where the measurement and force loops are coincident. Here it is through the stylus arm and specimen. Beyond these the two loops diverge through different components of the instrument. Forces will have minimal influence on those parts of the loops. In the alternative design of Figure 3.14 the force and measurement loops are coincident and there is common path with the drive forces through the bearings. This identifies the design as inferior (from this point of view, at least) even if the loops were the same physical size. This example illustrates another general rule in design:

in any instrument or machine, measurement and force loops should be separated as far as is possible.

The ultimate example of this principle is in high precision machine tools where the main structure is shadowed by an unloaded metrology frame from which all control datums are taken.

3.5 Principles of alignment

3.5.1 'Cosine' errors

If the displacement of a carriage is to be measured, or a specified displacement imposed upon it, it is important that the axis of the movement is parallel to the axis along which the measurement is made. If not, the carriage displacement

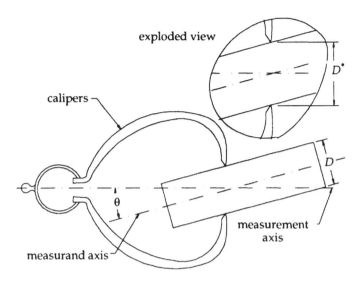

Figure 3.16 Alignment error in diameter measurement

will be the projection of the measured value onto its own axis, so the measurand exaggerates the true value. Variations on this theme occur very commonly and can never be totally eliminated.

A basic form of this alignment error is illustrated in Figure 3.16 where a round rod is being measured using a pair of callipers which are inclined at an angle θ relative to its axis. The error between the measured value, D^*, and 'true' value, D, is by simple trigonometry

$$D^* - D = D^*(1 - \cos\theta) \qquad (3.7)$$

As it occurs so often, this is often referred to as the 'cosine error'. Beware that this expression is also often used dismissively, with the implication that it is not worth worrying about. While it is true that cosine varies from unity only slowly with small angles, that variation could become significant in high range to resolution applications. A misalignment of $1°$ produces a cosine error of 0.015%, $5°$ one of 0.4% and $10°$ one of 1.5%. For $\theta \ll 1$ rad, the error is well approximated by $D^*\theta^2/2$. Since one cannot expect the angular misalignment to be constant during real operation, a corollary of Equation 3.7 is that it is poor practice to use a long probe between a sensor and a moving point if a short range, high precision measurement is wanted.

3.5.2 Abbe offset

Many of the good practical rules for precision engineering were first expressed by the nineteenth century optical instrument makers and astronomers. Ernst Abbe, founder of the Carl Zeiss Foundation, made several contributions of which the most telling is his alignment principle. It is simple to state, but its implications are legion and of great practical importance:

> *when measuring the displacement of a specified point, it is not sufficient to have the axis of the probe parallel to the direction of motion, the axis should also be aligned with (pass through) the point.*

The reasoning behind this is shown in Figure 3.17. A probe monitors the displacement of point P, guided by its own bearing axis, AA. This axis is a perpendicular distance a, often called the Abbe Offset, from the line of motion of P. Providing that the two axes are exactly parallel, the change in probe reading, s, correctly records the displacement of the point, d. However, if the axes becomes misaligned by an angle θ an apparent change in position will be recorded even though P has not moved. If the real position relative to the probe reference point is d_0, the apparent movement, s_θ will be

$$s_\theta = \frac{d_0(1 - \cos\theta)}{\cos\theta} - a\sin\theta \qquad (3.8)$$

The first term is a typical cosine error, which may well be negligible, but the second term, the Abbe error, could be much more significant since it is directly proportional to the misalignment at small angles. Abbe's principle asserts that a should be zero, that is s_θ is minimum according to Equation 3.8.

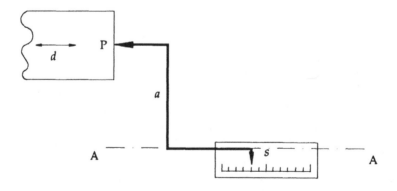

Figure 3.17 Measuring displacement with an Abbe offset.

If the angle of misalignment is finite but fixed, then the error in s of a positional change, d, depends simply on the cosine term. There is no incremental effect in the Abbe error, so in principle it could be calibrated out. The danger arises because as the probe follows the point P it moves relative to its own bearings, resulting inevitably in small variations in the angular alignment. A large Abbe offset will impose these variations onto the measurement at much higher significance than will the cosine term.

Abbe's principle was originally, and is usually, expressed in terms of measurement, but it is obvious that it applies also to the imposition of a displacement by an actuator. For best precision the drive axis should be aligned with the datum point of the carriage. A typical example of offset might be on a lathe where gauging is applied along the axis of the lead screw, whereas the tool is mounted to a post above the saddle, yet it is really the position of the tool that should be controlled. Sometimes it is difficult to avoid some offset, but it is common to see unnecessarily large ones caused through a lack of forethought. Several variations of this principle will be observed in other parts of this book and sometimes design guidelines seem more logical if this underlying concept is kept firmly in mind.

3.6 Nulling

So far, it has been implicitly assumed that the characteristics of instrument designs are sufficiently linear and repeatable that variations in their output can be interpreted directly in terms of a measurand such as displacement. To do this springs, transducers and other components of the system must remain sufficiently constant over time and environmental change and not be affected by the measurement itself. Achieving these conditions becomes increasingly difficult as sensitivities increase and alternative methods are sought. Nulling, a special case of feedback, is one such. The high sensitivity of a nonlinear or rather variable phenomenon is exploited by using it as a sensor in a system that is driven to maintain a constant output value compared to some reference. The drive signal is taken as the output measurement. The detailed characteristics of the sensor are unimportant since only a single point in its operating range is used. One familiar example is the assay balance (discussed further in Chapter 5). The mass to be measured is 'balanced' against known, reference masses. Its essential feature is that the measurement is taken only when the balance arm has been brought back to its original position, thereby eliminating potential errors from nonlinearities of the knife edge pivot about which it rotates.

The tunnelling probe for the measurement of surfaces provides an excellent, more recent application of nulling. The principle of quantum electron tunnelling is relatively straight forward and is illustrated in Figure 3.18. A

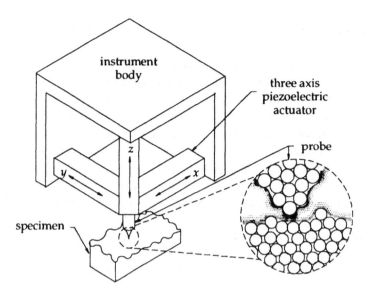

Figure 3.18 The scanning tunnelling microscope.

tunnelling probe consists of a very sharp electrode that is positioned near to an electrically conductive surface under investigation. The electrons in both probe and specimen are not absolutely bound by their respective surfaces but have an exponentially decaying probability of extending some distance beyond them. Consequently, the application of a small electrical potential between the probe and specimen will result in a finite current even though the gap represents a classical potential barrier. As the probe moves closer to the specimen then the gradient of electron density induces a flow of electrons (a current) that is proportional to the potential difference and strongly (exponentially) dependent upon the probe to surface separation – a more detailed discussion is presented in Section 7.3.4. Many other parameters such as the work function (the amount of energy necessary to remove an electron from the surface) and the amount of energy necessary to bridge the gap (related to dielectric strength) make it very difficult accurately to assess the current-separation characteristic in practice. It will depend upon both surfaces and environmental conditions.

The usefulness of the tunnelling probe rests on the extremely strong relationship between current and separation. Typically the tunnelling characteristics only extend to separations of up to 2 nm. If a constant potential is applied to a tunnelling probe and a relatively constant current maintained, the probe to specimen separation must be being held constant to considerably better than this value. A surface profile is obtained by moving one of the electrodes (usually the probe) relative to the other while maintaining a constant separation. The 'vertical' motion of the probe during a scan is taken to represent the surface profile. Effectively, we have a 'follower' technique, similar to using a minute stylus, except that it may be considered non-contact. The profiling accuracy depends on knowledge of the height control servo, but is almost independent of the actual tunnelling condition at the tip. It is analogous to measuring the altitude, relative to a fixed reference, of a hovercraft that closely follows the profile of the ground over which it is travelling.

Typically motion of the probe is achieved by mounting it onto a system of piezoelectric elements, for example the orthogonal set shown in Figure 3.18 although that geometry is rarely used in real devices. Each actuator is assumed to extend proportionally to an applied electric field (for a more complete discussion see the following section). An image is obtained by raster scanning the tip with the x and y axis actuators while the z axis actuator is servo-controlled to maintain constant separation. Historically, the principle of scanning tip profilometry was first realized by Young *et al.*, 1972, further studied by Binnig *et al.*, 1982, and shortly refined into the scanning tunnelling microscope (STM), Binnig and Rohrer, 1982, and another related technique, the atomic force microscope (AFM), that senses van der Waals forces or repulsive forces across small tip to specimen gaps, Binnig *et al.*, 1986. Both consist of a very sharp proximity sensor that is advanced towards the specimen surface until a sufficiently strong signal is monitored, in the latter case by monitoring deflection of a beam upon which the probe is mounted. Atomic scale lateral resolution can be achieved and vertical sensitivity is to a small fraction of atomic spacings. Lower but still very good resolution is achieved with other sensors used in similar ways; resonant beam, magnetic forces, capacitance, thermal, optical, acoustic. For a review see Kuk and Silverman, 1986, Chetwynd and Smith, 1991. In all cases, the characteristics of the probe-surface interaction are highly nonlinear and often unpredictable but, to reiterate, errors are reduced by operating about a null position. Dependence on the sensing characteristic can never be totally eliminated, however. This can be illustrated with the STM by varying the applied tip potential during imaging, when an apparent change in the height of the surface and slight morphological changes, at the atomic level, of the profile will be observed, Moller *et al.*, 1990. Because of the extreme separation sensitivity of tunnelling, atomic resolution is nevertheless maintained.

3.7 Compensation

We saw in the previous section that tunnelling probe current (often in the region of 50 pA to 10 nA with an applied potential from 10 mV to a few volts) has an exponential dependence on the tip to specimen separation. Although excellent for detecting the presence of the specimen surface, this characteristic causes difficulties for the design of control circuits to position the probe. Ideally a linear signal is required. The tunnelling relationships cannot be changed, so usually the detector signal is passed through a logarithmic amplifier before use in the controller. The undesirable characteristic is made more tractable by multiplication by an inverse form. In other situations an equal and opposite effect might be added. They are examples of *compensation*. Electronic compensation has become extremely important with the ever increasing complexity of integrated circuit devices and provides a powerful weapon in the battle for better performance. This book, however, is concerned primarily with mechanical design and, although discrimination between the disciplines of mechanics, electronics and physics is becoming increasingly blurred, we concentrate on ideas commonly employed by the mechanical designer. Digital computer based compensation strategies will also be addressed briefly.

3.7.1 Mechanical compensation strategies

The demand for navigational chronometers, both clocks and watches, in the seventeenth and eighteenth centuries marks an important point in the evolution of modern precision engineering. Watch makers soon realized a need to compensate for thermal expansion of materials. Thermal expansion of the pendulums used as a timing mechanism since Galileo was identified as a major contributor to timing errors. A large number of ingenious mechanisms were used to compensate for this effect, for example Arnold's double-S compensated balance described in Smith, 1975. Two other more direct devices are shown in Figure 3.19.

Figure 3.19a shows a flywheel for use as the inertial mechanism (balance wheel) in the spring-inertia oscillator of a pocket watch. Thermal expansion is compensated by distortion of the bimetallic strip formed by using two materials having different thermal expansivities. Both the size and shape of the wheel change with temperature, but in such a way that the radius of gyration remains almost unchanged. Clearly material 2 must have lower thermal expansivity than material 1. In practice it is rather difficult to set the ratio precisely but improvements over a simple wheel are readily obtained. Figure 3.19b shows a rigid pendulum rod, light compared to the mass on its end. The rod is constructed from two lengths of different materials. One has a negative thermal expansion coefficient and the other a positive one. The rod will maintain a constant length, if the ratio of thermal coefficients is equal to the inverse ratio of the corresponding lengths. Negative expansivity materials are rare, so alternative methods of simulating this situation have been devised, for example as serpentine pendulums in which 'upwards' and 'downwards'

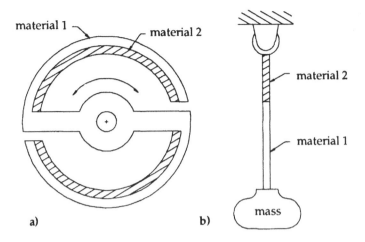

Figure 3.19 Examples of constant frequency oscillators for clocks and watches; a) constant inertia, b) constant length.

expansions of rods linked at their ends leads to a near zero change in the overall inertia.

An alternative approach is to seek materials of such small expansivity that compensation is unnecessary. One quite early discovery was the 36% Nickel/Iron alloy often called Invar. Its close relative Elinvar (42%Ni) has a low thermoelastic coefficient which is also important for watches since both the spring and the inertia must remain constant (or at least track each other's changes). More recently the ultra-low expansion glass ceramics have provided orders of magnitude improvements in thermal stability. The internal mechanism is one in which different material phases of the ceramic compensate thermally on a microscopic scale.

Looking back at the example of the stylus instrument, Section 3.4, the large number of components may seem to work against thermal stability. However, performance depends only on the stability of the *metrology loop*. Often, as in this example, it is only in certain directions that high stability is needed and then constructing all components from the same material may suffice. An expansion of one component in the loop will be compensated by an equal and opposite expansion of components in an opposite leg. If a small air-gap is needed, a short brass insert in the same leg of an otherwise steel loop will easily supply the extra expansion of the 'missing' steel. More sophisticated uses of mixed materials are readily developed. For *changes* in *uniform* temperature, a system constructed from similar materials may be close to self compensating. Under dynamically varying thermal conditions compensation is much more difficult since it depends, for example on the thermal capacity of loop components and materials selection must include dynamic thermal effects (see Chapter 8, particularly, Section 8.3.2).

Along with loops, actuators commonly incorporate compensation principles. Feedscrews will be used to illustrate. A long screw will have some slow variation in its pitch. This arises from its manufacture and mounting and so will be systematic. A perfect nut will transmit these errors into the machine motion. We note, however, that if instead of it being locked, a rotary motion is imposed directly on the nut then the screw and nut together form a differential mechanism. In principle the nut could be wobbled about its nominal position such that the differential motion just compensates for the screw error. Successful mechanical compensators have been built in which the rotary motion of the nut is controlled by a rigidly attached torque-arm that runs along a linear cam. The cam must be made to match the errors of a specific screw after installation, so this is an expensive undertaking. A feedscrew will also have short range irregularities of the thread form that cannot readily be compensated by this method. In a perfectly made, rigid, screw and nut, the two surfaces will conform to each other with no clearance to provide a precise, high stiffness drive. Since errors are always present, it is necessary to use a compliant nut to achieve this high degree of conformity. In other words, elastic averaging is used both to improve the definition of the motion and to increase the effective stiffness. This was recognized by designers of early diffraction grating ruling engines, used to scribe straight, parallel lines, up to 800 per mm, onto flat or cylindrical surfaces, see for example Merton, 1950, Hall and Sayce, 1952, Stanley et al., 1968. Henry Rowlands in the 1870s used a feedscrew and nut arrangement with the nut and the slideway being produced from the oily hardwood lignum vitae. Others used cork, leather and even the pith of an elder tree to mechanically average the short range feedscrew errors. Modern designers make use of PTFE to achieve linear relationships of better than 1 part in 10^6 between rotation of the screw and motion of the nut (including the longer range compensation).

While sensor nonlinearities can often be sidestepped through the use of nulling (Section 3.6), actuators must be used across their range and so nonlinearity must be compensated actively. Feedback drives utilizing sensors of adequate linearity are often used, both on large machine tools and in microactuators. Piezoelectric materials are much used for short range applications such as the scanning tip microscopies discussed earlier and in the previous section. The assumption that their extension is directly proportional to applied field is far from true in reality. They exhibit a large hysteresis and the extension will also strongly depend upon both temperature and time, see Section 7.1.1. These limitations were overcome by Hicks et al., 1984, when displacing and maintaining the parallelism of plane mirrors for optical interferometry in a Fabry-Perot etalon. The mirrors were both spaced apart and moved by a series of piezoelectric actuators about their periphery, but the position was monitored by capacitance sensors directly deposited on the mirror surfaces. The capacitors provided the highly precise positional references for the control in which, with heavy feedback, the forward loop characteristic of the actuator had little direct influence.

This idea has been applied in several commercial piezoelectric actuators. A cross-section of a typical actuator is shown in Figure 3.20, where a stack of disc piezoelectric elements is driven by a high voltage amplifier. Each end of this stack is connected to an Invar cap, to which external devices may be attached. An ultra-low expansion glass-ceramic ('Zerodur') rod is positioned in the centre of the tube, attached to one cap. An electrode deposited on its free end forms part of a capacitance gauge that monitors the extension between the two end plates. The measurement loop in this design passes through the centre of this rod and gauge and into the end caps. However, the force loop passes through the piezo-stacks and the two loops only coincide at the end caps. Both loops then pass around the structure to which it is attached and which is beyond the control of the actuator designer. The internal separation and common external path provides the best compromise for this situation. Any error between the position sensor and a specified reference value (or demand)

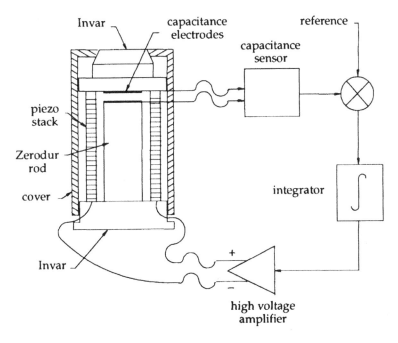

Figure 3.20 Piezoelectric actuator with inbuilt capacitance position sensing (reproduced with kind permission of Queensgate Instruments).

is used to provide a drive signal to the actuator via an integrator. The integrator is necessary because the field applied to the piezostack must be maintained if it is to hold a constant extension. If a simple proportional signal were used, the drive signal would go to zero as the actuation error decreased and the device would relax. Another advantage with an integrating control is that if, for example, a load is applied to the actuator resulting in a distortion monitored by the gauging, the integrator will continue increasing the drive over time until its fully compensates the force and returns the capacitance electrodes to the desired separation. It is usual to combine proportional and integral control for an optimum response, with derivative actions often added to reduce transient overshoots by simulating dampers without actual energy dissipation. A discussion of controller strategies is beyond the scope of this book: good introductions to this important topic can be found in, for example, Richards, 1979, Dorf, 1980, or Warwick, 1989. The relative position of the two end plates can be set with sub-nanometre precision over a range of 15 μm under applied loads of up to 10 N or more and with a temperature coefficient of only about 15 nm K^{-1}. Note that a solid rod of Invar of similar overall dimensions would have an expansion of around 45 nm K^{-1}. The potential of this type of device for displacement transfer standards of sub-nanometre accuracy has been explored by Harb *et al.*, 1992.

The piezoelectric actuators in the STM of Figure 3.18 are used without compensation. This is true of practically all the development of these microscopes over the last fifteen years. For studies of surface physics on regular structures post-measurement image correction could be applied, so there was little early incentive to increase system complexity. For many potential applications, positional uncertainty may be a serious limitation and some recent workers have tried to compensate by incorporating laser measurement systems into each axis, Tsuda and Yamada, 1989, Barrett and Quate, 1991.

3.7.2 Computer compensation of systematic errors

Systematic errors can be compensated ('corrected' is often used, but tempts over-optimism) because they repeat. Mechanisms that record errors and subsequently imposed corrective motions, such as the pitch correction cam discussed in Section 3.7.1, are expensive to produce. The obvious method of storing and reproducing the errors is now to use a digital computer. It may well prove economic to install a microcomputer just to improve working precision, but so many machines and instruments now use computer control and analysis, that the costs of the compensation are often only marginal. Consider, for example, a simple fine angular drive consisting of a motor driven micrometer pushing against a tangent arm lever. The actual angular change caused by some fixed increment of the micrometer will vary with the angle of the lever to the line of action of the micrometer: the drive is non-linear. This error is geometrically predictable and could be compensated by profiling the lever suitably where it contacts the micrometer. However it is probably less expensive to use a simple mechanical form and to incorporate a correction term, computed from either a formula or a look-up table, into the drive motor control. It may well be the case that the best strategy for designing a precise mechanical movement involves selecting for repeatability even at the expense of path accuracy and then incorporating a non-linear drive control.

The case for computer compensation is even stronger when the systematic error cannot be predicted at the design stage since it is likely to have an irregular form that is difficult to reproduce mechanically. Since the error must be measured it might just as well be fed straight into a computer memory. The classic case of this approach is with machine tool numerical controllers, where the errors in the positional control are reduced by interpolating between a set points calibrated from a laser interferometer during the machine installation. This approach can be extended to several axes and to provide closer spaced set points for better precision, until ultimately a complete calibration net of the working space of the machine is created. Even with modern high speed, low cost computers, the amount of data that must be collected for this process rapidly grows out of control and in practice a relatively small set of points is used. This approach has been applied to coordinate measuring machines, for example Burdekin *et al.*, 1985, Busch *et al.*, 1985, where the complexity is soon realized. Each axis can display errors in each of its six degrees of freedom and there are also interactions between the alignments of the axes, so a simple three axis machine has 21 independent error profiles that combine to cause the overall volumetric uncertainty.

A rather different aspect of compensation is seen in instruments that measure shape, that is they record the deviation of the profile of a workpiece from that of a perfect geometric figure. Real measurements will clearly include contributions from the errors in the machine datums as well as those of the workpiece. The only sensible definition for the machine error is the shape of the profile that would be obtained if a perfect workpiece were to be measured. Thus the obvious difficulty in determining the errors is that of obtaining a perfect workpiece, or, at least, one that is known to be an order of magnitude nearer perfection than the machine datum. This problem is circumvented by the trick of considering a real measurement as, in turn, that of the workpiece corrupted by the datum error and of the datum corrupted by the workpiece error. Both of these errors are systematic, but the way in which they combine depends on the relative orientation of the workpiece and datum. Hence measurements made in different relative alignment will yield different net profiles. Knowing that these profiles are constructed from the same pair of underlying profiles and knowing also the shift in alignment, it may be possible to determine which belongs to the datum and which to the workpiece. Such techniques are known as error separation.

The simplest error separation technique is known as reversal and works by realigning the workpiece between two measurements so that the direction of the workpiece profile relative to the datum is exactly reversed. The two measured profiles can then be taken to be the sum and difference of the individual ones and separation is quite straightforward. This has been applied to the measurement of nominally straight sections, Thwaite, 1973, and to circular ones, Donaldson, 1972, Chetwynd, 1987. Roundness measurement is the easier to visualize. Figure 3.21a shows schematically a nominally round workpiece placed on a precision turntable. A linear displacement probe, acting radially to the centre of rotation, contacts its surface and so records the profile deviations. A dot on the table and the part show their relative orientation. We assume that the workpiece has an out of roundness profile $w(\varphi)$ and that the turntable spindle has an error, that is axial shifts in the line of the sensor as it rotates, $s(\theta)$. Initially the two systems are aligned, so $\varphi = \theta$. Now assume that the workpiece is exactly centred to the axis of rotation (of course, this is impossible in practice, but there are excellent ways of computer compensating for the error caused by small eccentricites that will not be considered here). The recorded profile will be simply

$$p_1(\theta) \;=\; w(\theta) \,+\, s(\theta) \tag{3.9}$$

Now rearrange the set-up to that of Figure 3.21b. The workpiece has been rotated by 180° with respect to the turntable, as has the probe. Thus synchronism between the workpiece profile and the probe signal is still the same, so still $\varphi = \theta$. However, if at some point in its rotation the spindle axis shifted slightly to the right, the original set-up would see an outward radial movement, as if there were a bump on the workpiece, while in the latter set-up that same shift would be seen as radially inwards, equivalent to a hollow in the workpiece, since the probe has been repositioned. The overall effect of the changes in the set up is to reverse the sign of the spindle error so that the recorded profile will be

$$p_2(\theta) \;=\; w(\theta) \,-\, s(\theta) \tag{3.10}$$

Adding the profiles of Equations 3.9 and 3.10 yields $2w(\theta)$, while subtracting them gives $2s(\theta)$. Performing the pair of measurements allows theoretically complete separation of the errors. Once this is done the spindle error can be stored and simply subtracted from subsequent measurements. Since the systematic spindle error may change in the longer term because of environmental changes or running-in of the bearings, better overall accuracy

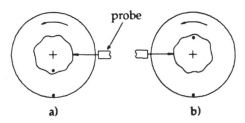

Figure 3.21 Schematic representation of error separation by reversal in roundness measurement.

will be obtained by using the reversal process directly to extract the workpiece error. This is economically feasible for many super-precise measurements since the change between settings is simple to execute. Perfect separation of errors is not obtained in practice because the reversal itself cannot be executed perfectly. There are also some subtleties concerning the definition of errors that need not be explored here. Chetwynd, 1987 discusses such points in some detail.

Good quality error separation does not always require reversal. In the case of roundness measurement, both the workpiece profile and the spindle error must be periodic with the rotation. They may therefore be described by Fourier series. A series of measurements, all with the probe set as in Figure 3.21a, is taken in which the orientation of the workpiece to the table (that is, $\varphi - \theta$) is incremented successively by some known phase angle. Mathematical best fits of different harmonics in these profiles then identify how much their phase is advancing and so the relative magnitudes of the fixed phase signal (spindle) to the moving phase one (workpiece), Whitehouse, 1976. Complete separation is not achieved since the number of harmonics that can be included depends upon the number of independent measurements made. The method may be easier to automate since the probe does not move and a simple indexing table suffices to adjust the workpiece. The indexing and reversal approaches both succeed reliably in reducing the effect of systematic spindle error to as little as one tenth of its actual magnitude, Chetwynd and Siddall, 1976.

3.8 Symmetry

Previous sections of this chapter have discussed loops and alignment, feedback, nulling and compensation. The examples shown indicate that there is a strong interaction between such notions in the design process and so we look for underlying concepts that may help to unify them. Undoubtedly symmetry is an important one, which we stress by giving it this short section to itself. A typical dictionary definition of symmetry is:

the state in which one part exactly corresponds to another in size, shape and position; harmony or adaption of parts to each other.

This definition speaks for itself. Even the more romantic idea of harmony is relevant in the reassurance one feels, when looking at a symmetric structure, that it is a 'good' design. This is partly aesthetic but there also many technological advantages. If components within a measurement loop are made from the same material then thermal expansion related changes around that loop will probably compensate. This notion is just a form of the symmetry argument. It derives from the observation that if the components on the opposite sides of the loop are identical then *all* properties will be compensated both statically and dynamically. This would include electrical, optical and mechanical behaviour. The idea of mixing materials to achieve thermal compensation can be thought of as introducing thermal symmetry by the provision of additional mechanical asymmetries in a system that cannot be totally symmetrical. If a structure must carry a heat source and act as part of a measurement loop, it is a good idea to place that source at an axis of symmetry. Then equal conduction paths and thermal capacities should lead to equal thermal gradients and so to minimum disturbance of the loop. Note that simple mixed material compensation normally works only for changes in uniform temperature, not for gradients or transients.

Examples of thermal symmetry have been used because temperature stability is often a particularly challenging aspect of precision design. It is by no means the only important one. Mechanical distortion of loops tends to occur most seriously from bending stresses in their elements. Again, good practice suggests that the best performance will be obtained from a metrology loop if the measurement (Abbe) axis lies on an axis of symmetry of the stress distribution. The assay balance is an excellent example of this principle. When in the null condition it has a high degree of symmetry about its pivot with lengths and shapes, positions, and forces and stresses being *balanced*. It is immune to almost every uniform environmental influence and tolerant of many non-uniform ones. Many other examples of both mirror and circular symmetry will be found when perusing the pages of this book and other books and journal articles in which instrument designs are presented.

3.9 The effects of scaling: Stiffness and dynamic response

The revolution in microelectronics has brought with it the possibility of producing similarly miniaturized mechanical devices. Although related work had been going on for some years previously, the (still very useful) paper by

Petersen, 1982, on the use of silicon processing technologies for the production of mechanical devices marked a rapid growth of interest in the manufacture and application of them. The first practical production devices were primarily diaphragm-based pressure sensors, beam resonators for accelerometers and, a little later, nozzles for ink jet printers. They were chosen for their simplicity and the fact that they lend themselves readily to planar production, already well developed for microelectronics manufacture. The principal techniques rely principally on the deposition of materials in thin layers and lithographic etching. A mechanical device is built up in several stages at each of which material is either deposited, usually from a chemically active gaseous phase, or chemically etched through a photoresist mask to modify the structure. Sometimes energy beams, ions or electrons, are used for material removal.

To illustrate the process, we outline the process described by Tai and Muller, 1989, for the production of the rotor of an electric motor, see Figure 3.22.

First, a layer of silicon nitride on a thinner one of silicon dioxide is deposited onto a substrate material. This provides a chemically resistive, etch-stopping layer that forms the base plane. A ground plane of polycrystalline silicon for the rotor is then deposited on top, see Figure 3.22a. Figure 3.22b shows the situation

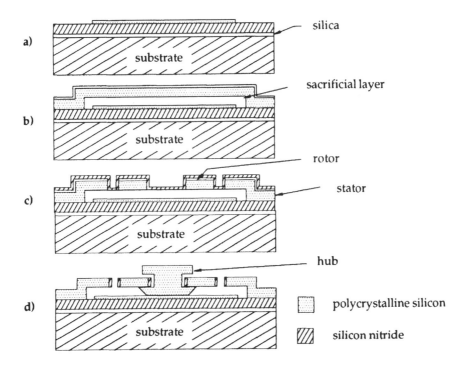

Figure 3.22 Various stages for the production of a rotor of a micromotor.

after successive layers of phosphosilicate glass (PSG), polysilicon and silicon dioxide are deposited. The glass provides a sacrificial layer that can later be removed without affecting significantly the silicon structures formed around it. The next stages involve plasma etching, low pressure chemical vapour deposition (LPCVD) and reactive ion etching through various masks to produce separate, nearly complete stator and rotor structures on and around the glass, protected by thin nitride coatings, Figure 3.17c. Finally, a further sequence of etching and deposition processes 'drills' through the central area of the PSG to the lower silicon layer, grows the silicon hub, which is then etched back a little to give clearance, and then removal of the rest of the PSG leaves the complete motor of Figure 3.17d. Typical complete motors are no more than 200 μm across and a few tens of micrometres deep. Silicon and associated metal growth technology is now producing a wide variety of devices including tweezers, gyroscopes and optical filters. The list is growing almost monthly.

Why produce miniature components? As with microelectronics, there are some applications where size is itself the limiting factor, but more commonly the reason lies in the way performance changes as the scale is reduced. Miniaturization leads to improved dynamic response. Without attempting a full analysis of motor dynamics, it is clear that its natural frequency places an upper bound on its response. It is therefore sufficient for present purposes to study a simple model; a thin torsional rod that is rigidly fixed at one end and attached to a rigid mass at the other, see Figure 3.23a.

If the mass of the support rod and compliance of the disc are neglected, Figure 3.23a represents a simple harmonic oscillator of natural frequency

$$\omega_n = \left(\frac{\lambda_\theta}{I}\right)^{1/2} \tag{3.11}$$

where λ_θ is the torsional stiffness of the rod and I is the moment of inertia of the disc. If both are cylindrical homogeneous, isotropic solids of density ρ and modulus of rigidity G, then, for the parameters shown on the figure,

$$I = \frac{1}{2} mr^2 = \frac{1}{2} \rho \pi r^4 h \tag{3.12}$$

$$\lambda_\theta = \frac{\pi d^4 G}{32L} \tag{3.13}$$

and substituting Equations 3.12 and 3.13 into Equation 3.11 yields

$$\omega_n = \left(\frac{d^4 G}{16 \rho r^4 L h}\right)^{1/2} \tag{3.14}$$

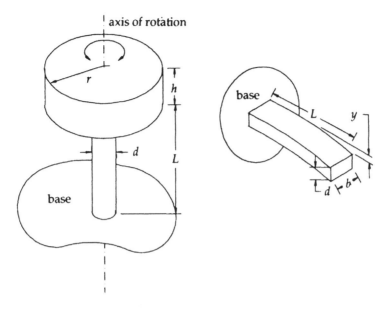

Figure 3.23 Simple solid structures; a) torsional oscillator, b) a cantilever beam distorting under its own self weight.

To investigate the effect of scaling on the natural frequency, we introduce a scale factor, k, as a multiplier of all spatial dimensions. We assume that the density and modulus of rigidity are constant with scaling. This is obviously not true at the scale of atoms, or even of grain sizes in multicrystalline materials, when anisotropic behaviour predominates. Neither is it true with bodies as massive as black holes, but these are rare in precision engineering. All the same the assumption is reasonable to quite small sizes in most materials and to very small sizes in amorphous and single crystal structures. Counting the powers of dimensional terms in Equation 3.14, we see that the natural frequency of a scaled oscillator, ω_k, will be

$$\omega_k = \frac{\omega_n}{k} \tag{3.15}$$

This simple model indicates that the improvement in the dynamic response of any linear system is *equal to* the inverse of the scale factor: the smaller a system becomes, the faster it will operate. This does not necessarily mean that its efficiency will increase nor that it can carry out a prespecified task faster. It means only that the tasks that it is capable of performing can be executed more rapidly than with a larger device.

If improved dynamic response through scaling is to be exploited in practice, structures must also become stiffer to resist the body forces from their own mass as size is reduced. That this is so can be illustrated by considering a rectangular, prismatic beam cantilevered from a rigid wall, Figure 3.23b. From the analysis in Chapter 2, a horizontal beam of second moment of area I_a, sags under the effect of its total weight, W, and the maximum deflection, occurring at the free end, is

$$y_{max} = \frac{WL^3}{8EI_a} = \frac{3L^4 g\rho}{2d^2 E} \tag{3.16}$$

Again applying a scale factor to every spatial dimension reveals that y_{max} is directly proportional to the square of the scale factor. The absolute deflection decreases rapidly as size is reduced. However, the length of the beam is reducing linearly with scale factor, so the relative deflection is proportional to the scale factor. Smaller systems do indeed offer more resistance to body forces. A more graphic illustration might consider the contrasting effects of body forces on a spider and on an elephant as each falls down a mine shaft.

Two theoretical aspects of the benefits of scaling have been explored, but there are several others of practical significance. Here we merely state without further comment that other advantages of small scale batch fabrication of engineering mechanisms include the following.

1. As the size of components reduces so too does the size of (surface) cracks present in them. This leads to an increase in the strength of brittle materials thus making them more suitable for safety critical applications. Devices also become more load efficient in the sense that the possible work per unit volume increases.

2. Small devices have increased thermal response and increased sensitivity to many stimuli, so more effective transducers, notably sensors, may be produced.

3. The processing technologies for large scale integration of microcircuits enable also the low cost mass production of large numbers of devices of almost identical performance. This has considerable importance, for example, to the development of sensors for automobile applications. In the longer term, large squads of simple micro-robots might be used effectively to perform complex operations in hazardous environments, or in medical work. Drexler, 1990, speculates, amongst other things, that molecular sized devices might self-assemble or even model biological reproduction, leading to 'assemblers' for the microscopic building of large scale structures. We leave further thoughts on such topics to him.

3.10 Case study: A proposed nanohardness indentation instrument

Most of the case studies presented relate to existing systems so that there can be practical evidence of their effectiveness. However, it would be incongruous to produce a book of this type without attempting any creativity and this is an appropriate place so to do. We examine the outline design for a novel, physically small nanohardness measuring device. This is particularly challenging because the objective is to impress a controlled force onto a specimen through a small indenter diamond and *simultaneously* measure displacement at the same position. The total depth of penetration and both the elastic recovery and the final permanent indent depth are of interest. The device is to be used to study the materials properties in the atypical surface structure and so must operate with indents confined totally within such layers. Consequently, total displacements are usually sub-micrometre and the corresponding imposed forces no more than millinewtons. Resolution to better than about 1 nm for displacement and 0.1 μN for force are needed. A possible instrument design will be examined in the light of the concepts presented in this chapter.

A proposed system is shown in cross-section in Figure 3.24. At its heart, an electromagnetic force actuator is used to provide the reference indentation force because it may be readily controlled electronically and because it is a non-contact device. Capacitive micrometry has been selected for the gauging because it is compact and relatively easily provides sub-nanometre precision. We assume, here, that these methods provide adequate resolution and range; for further details on them, and rival devices, see Chapter 7. We concentrate on the mechanical layout, where some constraints are immediately clear. There must be a stable (thermally and mechanically) measurement loop, as short as possible, from the specimen, through a capacitive gauge and back to the probe/indenter. The indentation force loop is from the magnet of the actuator, through the indenter and specimen and back to the coil of the actuator. It must have common path with the measurement loop since the probe is used for both purposes, but the loops should follow separate paths for as much of their lengths as possible. A properly designed force actuator is known to provide good linearity even in the presence of small axial movements of the coil relative to the magnet. Thus an outer frame, which need not be of expensively high stability, can carry the coil and be attached to the main structure close to the specimen to shunt the forces away from the metrology loop.

Even if the force could be set exactly at the actuator, the indentation force may be different, for example because of friction at the probe supports as it moves. Placing a force sensor immediately behind the tip would seriously reduce the dimensional stability of the measurement loop. Consequently, we

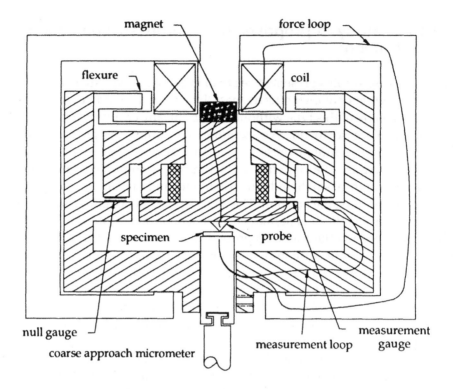

magnet

force loop

flexure

coil

specimen

probe

null gauge

coarse approach micrometer

measurement loop

measurement gauge

Figure 3.24 Schematic representation of proposed nanohardness instrument. The hatched sections carry the high stability loops, the cross-hatched component is a piezo-actuator.

choose to mount the probe on a flexure support that is theoretically frictionless and to avoid measuring the force directly. Deflection of the flexure will affect the net force at the indenter, but, unlike friction effects, it does so in a manner that is predictable and in principle compensatable in the drive. Even better, by incorporating a stiff, short-range length adjustment into the support, a nulling mode can be produced in which the springs maintain a constant deflection throughout the measurement cycle. The system then approaches the mythical ideal of a non-influencing support.

With these ideas in place, the layout of Figure 3.24 is quite easy to understand, and, indeed, to invent. The magnet is attached to the probe by a solid rod which hangs from a diaphragm linear spring flexure through a piezoelectric displacement actuator. The indentation depth measuring gauge is in parallel with the piezo-actuator. The flexure to actuator link is referenced to the rest of the metrology frame through a nulling gauge. If the signal from this gauge is kept constant through action of the piezo-actuator, the flexure is effectively removed from the measurement loop. Note, though that it is still necessary for the flexure to be totally within a high stability path, since distortion of its supports would affect the subsidiary measurement loop of the null gauge. The main loops are shown on the figure. It is tolerable for the null gauge to be within the measurement loop because its gap should be readily controllable to around 0.1 nm.

The next level of detail is best seen by considering the instrument operation sequence. First the specimen must be brought into proximity with the probe using a coarse infeed which is then decoupled to prevent the injection of noise. It must then be clamped firmly. Some fine positioning could be performed through the Poisson's ratio effect as the clamp tightens or by using a closed loop actuator similar to that in the upper half of the instrument. Now the force actuator can be energized. The force generated will pass through the probe stem to the flexures which will deflect slightly. Unless the probe is in contact with the specimen, this deflection will be monitored by the null gauge and feedback to the displacement piezo-actuator will advance the probe in an attempt to maintain a fixed gap at the null sensor. Upon contact with a specimen, the null signal will reverse in direction thus giving an almost instantaneous indication of contact. The uncertainty (from its ideal zero value) in the contact force at the initial detection of touching is simply the product of the stiffness of the spring and the resolution of the null gauging. If, for example, the flexure stiffness is around 1 kN m^{-1} and the gauge resolution better than 1 nm, the minimum force is below 1 µN which is approaching the resolution commonly required in nanohardness testing. However, being a null gauge, it is not difficult to scale the gauge sensitivity, by two orders of magnitude if necessary, by reducing its range commensurately. This is quite realistic since range is almost redundant in a null system. In fact, for an instrument of the size of a standard microscope, a stiffness of 1 kN m^{-1} is likely to cause unacceptable sensitivity to vibrations. Consequently, much of the extra sensitivity available by using the null gauging is likely to be used to allow a stiffer suspension without penalty to the force discrimination. Once contact is established, the application of force by the actuator will cause the probe to advance but deflection of the spring will be compensated by an expansion of the displacement actuator of equal magnitude to the depth of penetration of the probe into the specimen. This depth is monitored using the calibrated measurement gauge. The springs and the null gauge gap remain in fixed positions to a high degree of precision during this operation, so the full resolution of both the force actuator and the measuring gauge are manifest at the probe tip.

The design shown in Figure 3.24 can be criticized because the measurement gauge is laterally displaced from (although still centred around) the indenter rather than in ideal Abbe alignment. Consequently, errors from any tendency of the probe carriage to pitch or yaw as it moves will be magnified. An alternative configuration is shown in Figure 3.25. Both the indentation force and the indentation measurement are now in direct line with the probe. A potentially serious difference between the two designs is that the indentation force is now transmitted through the piezoelectric actuator in the measurement path between the gauge electrodes. However, no problems arise from its uncertain deflection due to this load because the null gauge is also in parallel with this section. The piezo-actuator will operate automatically to correct for its own finite compliance.

Note that its has been assumed, almost without thought, that the designs are symmetrical about the probing position. This is always a good idea in first designs: if it later becomes clear, for example, that more convenient access for specimen mounting requires that symmetry be broken, we can judge what performance might be lost. Some care must be taken in the interpretation of force and measurement loops. The force loop shown relates to the path taken by the imposed indentation forces. It does not take account of components that may induce forces if they distort. Here temperature stability is probably best achieved by using very low expansion materials for the measurement loops.

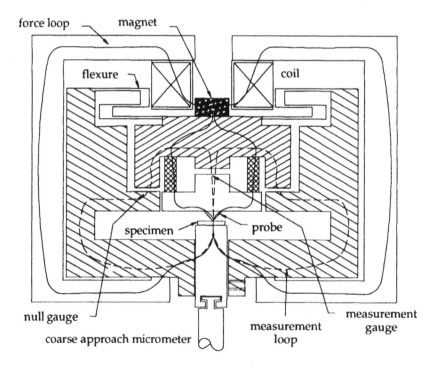

Figure 3.25 An alternative layout for the instrument of Figure 3.24, better complying with the Abbe principle.

Gravitational forces are constant when the instrument is held in one orientation and, in such a small system, are unlikely to cause any untoward loop distortion providing that sections thick compared to their length are used, as shown. If thermal distortion of the support flexure occurs it is likely to inject parasitic forces into the instrument loops. Such effects and mechanical ones like vibrations are impossible to eliminate totally. Either they are tolerated, or we enter the complex world of electronic or mechanical compensation and environmental temperature control that lies outside the scope of this discussion. Deciding on the best compromise is a perpetual, but exciting, challenge to design.

References

Archard J.F., 1957, Elastic deformation and the laws of friction, *Proc. Roy. Soc. Lond.*, **A238**, 190-205

Barrett R.C. and Quate C.F., 1991, Optical scan-correction system applied to atomic force microscopy, *Rev. Sci. Instrum.*, **62**(6), 1393-1399.

Binnig G. and Rohrer H., 1982, Scanning tunnelling microscopy, *Helv. Phys. Acta*, **55**, 726-735.

Binnig G., Rohrer H., Gerber Ch. and Weibel E., 1982, Tunnelling through a controllable vacuum gap, *Appl. Phys. Letts.*, **40**, 178-180.

Binnig G., Quate C.F. and Gerber Ch., 1986, Atomic force microscope, *Phys. Rev. Letts.*, **56**, 930-933.

Burdekin M., Di Giacomo B. and Xijing Z., 1985, Calibration software, an application to coordinate measuring machines, *Software for Coordinate Measuring Machines*, ed. Cox M.G. and Peggs G.N., 1-7, National Physical Laboratory, London.

Busch K., Kunzmann H. and Waldele F., 1985, Calibration of coordinate measuring machines, *Precision Engineering*, **7**, 139-144.

Chetwynd D.G., 1987, High-precision measurement of small balls, *J. Phys. E: Sci. Instrum.*, **20**, 1179-1187.

Chetwynd D.G. and Siddall G.J., 1976, Improving the accuracy of roundness measurement, *J. Phys. E: Sci. Instrum.*, **9**, 537-544.

Chetwynd D.G. and Smith S.T., 1991, High precision surface profilometry: From stylus to STM, in *From Instruments to Nanotechnology*, ed. Gardner J.W. and Hingle H.T., Gordon and Breach, London.

Chetwynd D.G., Liu X. and Smith S.T., 1992, Signal fidelity and tracking force in stylus profilometry, *Int. J. Mach. Tools Manufact.* **32**, 239-245.

Donaldson R.R., 1972, A simple method of separating spindle error from test ball roundness error, *C.I.R.P. Ann.*, **21**, 125-126

Dorf R.C., 1980, *Modern Control Systems*, Addison Wesley, London.

Drexler K.E., 1990, *Engines of Creation; A New Era in Nanotechnology*, Fourth Estate, London.

Hall R.G.N. and Sayce L.A., 1952, On the production of diffraction gratings. II. The generation of helical rulings and the preparation of plane gratings therefrom, *Proc. Roy. Soc. Lond.*, **A215**, 536-550.

Harb S., Smith S.T. and Chetwynd D.G., 1992, Sub-nanometre behaviour of capacitive feedback piezoelectric displacement actuators, *Rev. Sci. Instrum.*, **63**(2), 1680-1689

Hicks T. R., Reay N. K. and Atherton P. D., 1984, The application of capacitance micrometry to the control of Fabry-Perot etalons, *J. Phys. E: Sci. Instrum.*, **17**, 49-55.

Kuk Y. and Silverman P.J., 1986, Role of tip structure in scanning tunneling microscopy, *Appl. Phys. Letts.*, **48**(23), 1597-1599.

Liu X., Smith S.T. and Chetwynd D.G., 1992, Frictional forces between a diamond stylus and specimens at low load, *Wear*, **157**, 279-294.

Merton T., 1950, On the reproduction and ruling of diffraction gratings, *Proc. Roy. Soc. Lond.*, **A201**, 187-191.

Moller R., Baur C., Esslinger A., Graf U. and Kurz P., 1990, Voltage dependence of the morphology of the GaAs(110) surface observed by scanning tunnelling microscopy, *Nanotechnology*, **1**, 50-53.

Petersen K.E., 1982, Silicon as a mechanical material, *Proc. IEEE*, **70**(5), 420-456.

Phillips, J., 1984, *Freedom in Machinery*, Cambridge University Press.

Richards R.J., 1979, *An Introduction To Dynamics and Control*, Longman, London.

Slocum A.H., 1988, Kinematic couplings for precision fixturing-Part 1: Formulation of design parameters, *Precision Engineering*, **10**, 85-91.

Smith A., 1975, *Clocks and Watches*, Ebury Press, London.

Stanley V.W., Franks A. and Lindsey K., 1968, A simple ruling engine for x-ray gratings, *J. Phys. E: Sci. Instrum.*, **1**, 643-645.

Tai Y. and Muller R. S., 1989, IC-processed electrostatic micromotors, *Sensors and Actuators*, **20**, 49-55.

Thwaite E.G., 1973, A method of obtaining an error free reference line for the measurement of straightness, *Messtechnik*, **10**, 317-318.

Tsuda N. and Yamada H., 1989, STM, AFM and interferometers in NRLM, *Bulletin of NRLM*, **38**(4), 403-407.

Warwick K., 1989, *An Introduction to Control Systems*, Prentice Hall International, London.

Whitehead T.N., 1954, *The design and use of instruments and accurate mechanism: Underlying principles*, Dover Publications, New York.

Whitehouse D.J., 1976, Some theoretical aspects of error separation techniques in surface metrology, *J. Phys. E: Sci. Instrum.*, **9**, 531-536.

Young R., Ward J., and Scire F., 1972, The Topografiner: an instrument for measuring surface microtopography, *Rev. Sci. Instrum.*, **43**, 999-1011.

4 FLEXURE DESIGN FOR POSITIONING AND CONTROL

This chapter considers the application of bending and torsion of solid elements within the linear elastic range as a means of providing precise short-range motions. An initial analysis of basic beam bending sets the scene for a discussion of the general points in favour of and against elastic design. Linear mechanisms are then discussed in some detail. One degree of freedom motions based on leaf spring and notch hinge mechanisms of increasing symmetry are examined. A briefer treatment of angular motion flexures suffices. Finally an overview of the dynamical performance of linear mechanisms serves to show both that general approaches may be readily applied and that only specific solutions are worth pursuing. Case studies examine the use of monolithic linear spring mechanisms in very precise instrument applications.

4.1 A simple flexure design

Many applications require an accurate displacement or rotation about a known axis, over a limited range. Typical examples include dial indicators, precision balances, pendulums, tiltmeters, gravitymeters, micro-accelerometers and many manipulators used in optics, crystallography and such fields. A popular and effective method of achieving small controlled displacements is to apply a force to an elastic mechanism of known stiffness. This is a different concept to that considered in the previous chapter where load averaging and kinematic overconstraint is incorporated through the use of elastic, or compliant, elements within the instrument structure. Here kinematic principles of mechanism design are used directly.

The simplest flexure mechanism is the cantilever, see Figure 4.1. A load, W, is applied to the end of the beam and its subsequent displacement in the y direction is calculated from linear beam theory. Generally, the deflected shape of a (reasonably slender) transversely loaded beam may be calculated from its bending moments using

$$M = EI \frac{\mathrm{d}^2 y}{\mathrm{d}x^2} \tag{4.1}$$

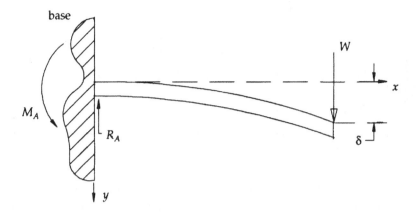

Figure 4.1   A simple cantilever beam

where E is the Young's modulus and I the second moment of area of the cross-section of the beam. Defining an origin at the built in end of the beam, Figure 4.1, and the x direction along its undeflected length, we find

$$M \;=\; M_A - R_A\, x \tag{4.2}$$

where M_A and R_A are the moment and force reactions at the origin. Integrating Equation 4.1 twice and noting that $\theta = y = 0$ at $x = 0$, $M_A = WL$ and $R_A = W$ for a beam of length L and constant I, the slope and deflection to the free end are

$$\theta \;=\; \frac{W}{EI}\left[Lx - \frac{x^2}{2} \right] \quad ; \quad \theta_{\text{end}} \;=\; \frac{WL^2}{2EI} \tag{4.3}$$

$$y \;=\; \frac{W}{EI}\left[\frac{Lx^2}{2} - \frac{x^3}{6} \right] \quad ; \quad \delta \;=\; \frac{WL^3}{3EI} \tag{4.4}$$

where δ is the deflection at the free end of the beam.

The simple cantilever is rarely usable because an arcuate locus is traced out by any point on the beam. To design a mechanism for linear or angular motion we use other geometries that exploit symmetry and superposition. A linear displacement at a point along the cantilever beam will be obtained if the slope there is consistently zero. In principle, this could be achieved by applying a couple to negate the twisting effect of the actuation load. One simple approach is shown in Figure 4.2, where the load is applied through a rigid bracket with its line of action s from the free end.

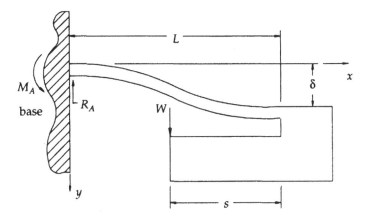

Figure 4.2 Deflection of a cantilever beam in the presence of an applied couple

The integration of the bending moment along the beam, from Equation 4.1, is exactly as in the previous case, except now $M_A = W(L - s)$ and $R_A = W$. Thus the slope and deflection are

$$\theta = \frac{W}{EI}\left[(L-s)\,x - \frac{x^2}{2}\right] \tag{4.5}$$

$$y = \frac{W}{EI}\left[\frac{(L-s)\,x^2}{2} - \frac{x^3}{6}\right] \tag{4.6}$$

We now observe, particularly, that if $s = L/2$

$$\theta_{\text{end}} = \frac{W}{EI}\left[\frac{L^2}{2} - \frac{L^2}{2}\right] = 0 \tag{4.7}$$

$$\delta = \frac{W}{EI}\left[\frac{L^3}{4} - \frac{L^3}{6}\right] = \frac{WL^3}{12EI} \tag{4.8}$$

To provide an end deflection that is free of rotation, the beam should be driven midway between the base and its end. This very simple design is not very stable because the beam is susceptible to buckling and, more importantly, twisting. Thus slight misalignment of the actuation force and any unexpected forces may cause rather large parasitic deflections. The torsional stiffness of a beam about the central, longitudinal, axis is given by

$$\frac{T}{\varphi} = \frac{GJ}{L} \tag{4.9}$$

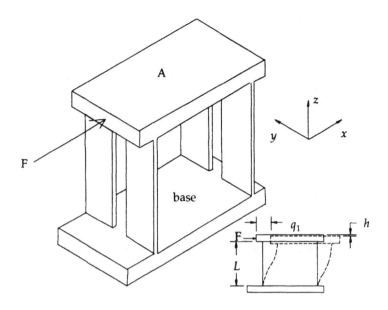

Figure 4.3   The simple leaf type linear spring; schematic representation of flexure distortion shown in side view (lower right)

where G is the modulus of rigidity and J is a constant based on the geometry of the beam cross-section, which becomes the polar second moment of area for circular shafts.

The resistance to torsional deflection is commonly improved by attaching two (or more) of these flexures together to form the simple linear spring mechanism as shown in Figure 4.3. Not only is this stiffer in torsion but it has the added advantage of providing a larger area on which to mount specimens or instrumentation. It is convenient, and therefore often done, to apply the actuation force to the platform, as shown. Equations 4.5 and 4.7 show that the tendency to pitch during motion is minimized if the actuation force is applied through the mid-point of the springs by means of a bracket attached to the platform. If the force is applied elsewhere, axial stresses develop in the springs to provide a moment that reduces the pitching, but such stresses are generally undesirable. This type of parallelogram motion spring is very common in instrument design and often provides the critical mechanism of precision instruments. For this reason, it will be analysed more deeply in Section 4.3 which considers linear mechanisms of high accuracy. First, however, it is useful to look in general terms at the advantages and disadvantages of flexure design.

4.2 The pros and cons of flexure design

Listing the merits and demerits of flexures, or any other devices, is necessarily a subjective process. What may be advantageous for one designer can be a distinct drawback to another. For example, for all the ease with which rectilinear mechanisms can be designed for specific applications, very few off-the-shelf devices are available. This might well be an economic disadvantage, though we tend to view it with favour since only a modest learning curve need be surmounted to use them effectively and the better understanding of principles so gained leads to better designs.

4.2.1 The advantages of flexures

1. They are wear free. This means that, provided they are not abusively distorted, their line of action (or position of the axis of rotation) will remain constant throughout their life. Because there are no sliding pairs, the only likely wear mechanisms will be fretting and corrosion at fabricated interfaces where changing stresses are applied. The rate at which the interfaces corrode or fret can increase if the interface consists of two different materials, Bowden and Tabor, 1964.

2. Flexures can be manufactured from a single piece of material to provide a monolithic mechanism which eliminates interface wear. Additionally, any instabilities introduced by either the high stresses associated with clamping or welding and the creep of glued joints are also removed.

3. Displacements are smooth and continuous at all levels.

4. It may be possible to construct devices that are insensitive to bulk temperature changes and even to temperature gradients in some planes.

5. Displacements can be accurately predicted from the application of known forces and, conversely, predictable forces can be generated by controlled displacements. If designed correctly, the springs will be closely linear in their force/displacement characteristics. In the inevitable presence of manufacturing errors, the displacement axis may not follow the ideal path, but it will still translate in a linear path.

6. Failure mechanisms due to fatigue or overloading of a brittle material are catastrophic and can be easily detected. Thus elastic-brittle flexures are especially attractive for systems that must have extremely high repeatability. For safety critical systems the more graceful degradation of elastic-plastic materials would be preferred.

4.2.2   The disadvantages of flexures

1. The force for a given deflection is dependent upon the elastic modulus of the flexure. It is surprisingly difficult to obtain values that are accurate to better than 1% for this parameter and it is often necessary to calibrate a flexure after it has been produced.

2. There will be hysteresis due to dislocation movement in most materials. Its magnitude depends upon the stress level, temperature and the grain structure and atomic bonding in the material.

3. Flexures are restricted to small displacements for a given size and stiffness. If stresses are to be kept low to ensure that the linear elastic region is being used, then the displacements must be correspondingly small.

4. Out of plane stiffness tends to be relatively low and drive direction stiffness quite high compared to other bearing systems. Great care must be taken to ensure that the drive axis is colinear with the desired motion. Methods for improving out of plane stiffness will be discussed in the following section.

5. They cannot tolerate large loads. The effective stiffness of a flexure system is reduced by the presence of externally applied loads. For example, there may be Euler (columnar) buckling if there is a load along a leaf spring. This is one of the main reasons that flexures are little used in machine tools and other applications where large and fluctuating loads will be encountered.

6. Unless precautions are taken, accidental overloads can lead to fatigue, work hardening and eventually to catastrophic failure. If a single, moderate overload causes some plastic deformation, the device may continue to operate as a linear mechanism, but with characteristics different to those prevailing prior to the overload.

4.3   Linear spring mechanisms

4.3.1   Simple and compound leaf spring mechanisms

Any mechanism in which displacements are achieved with little (ideally, zero) rotation will be considered as a linear mechanism. This should not be confused with rectilinear mechanisms in which we are looking specifically for motions in a straight line. The simplest linear mechanism is the cantilever beam of Figure 4.2, with s set to be $L/2$. Rearranging Equation 4.8 shows the stiffness, λ, of this mechanism to be

$$\lambda = \frac{12EI}{L^3} \tag{4.10}$$

If the spring has a rectangular cross-section,

$$I = \frac{bd^3}{12} \tag{4.11}$$

where b is the breadth and d the depth, in the direction of deflection. Thus some degree of tuning of stiffnesses in different planes is possible through the choice of b, d and L.

For the spring mechanism of Figure 4.3 the stiffnesses of the cantilevers add; its drive direction stiffness is twice the value given by Equation 4.10, that is $24EI/L^3$. Although the platform of the ideal mechanism will not rotate, it will follow a curvilinear path. The parasitic errors from rectilinearity have been studied by Jones, 1951 and 1956. He found that a simple spring mechanism of length 45 mm driven at the platform (A in Figure 4.3) through 1 mm gave a parasitic height variation of approximately 28.5 µm. This is a rather disappointing error ratio of 35:1 or 3% of traverse. There was a pitch of 30 seconds of arc with the two flexures spaced 25 mm apart. The ratio of the length of the supports to the distance between them, that is the slenderness of the flexure, was almost two, much higher than the values around unity that are commonly used. For small displacements there is an approximate equality between the ratio of length of flexure to deflection and the ratio of deflection to parasitic error. For example, for a deflection of 0.1 mm and a flexure length of 100 mm a change in height of 0.1 µm would be typical. Over very small displacements it is possible to achieve very accurate translations. Exploiting this phenomenon Jones, 1987, has been able to produce extremely sensitive instrumentation for measuring small variations in air pressure, gravity and even the tilt of Scotland as the tides rise higher on one side of the country than on the other! In recognition of his contribution, the leaf spring type of flexure is often referred to as the Jones spring. A more rigourous analysis of this structure is given by Plainevaux, 1956, who presents series expressions for the change in height of the platform, the stiffness and the bending moments about the ends of the flexure beams.

These mechanisms have the disadvantage that they have a relatively low stiffness in directions other than the drive axes. Rotation about the z axis of Figure 4.3 is usually the most compliant direction and thus most susceptible to parasitic errors in the drive mechanism. The common method for increasing the torsional stiffness without altering the drive stiffness is to split the leaf spring along its axis and to separate the parts as much as is practicable, see Figure 4.3.

This increases the effective polar second moment of area about the vertical axis of the system without increasing the planar one.

Manufacturing tolerances always degrade performance from the geometric ideal. For the fabricated Jones spring there are two sources of potentially significant errors. The first is a twist due to the difference, ε_b, in the separation between the flexures at the platform and at the base. This will result in a parasitic twisting about the y axis of magnitude

$$\theta \approx \frac{q_1^2 \varepsilon_b}{2bL^2} \tag{4.12}$$

where q_1 is the platform displacement, and b is the nominal horizontal separation of the flexures. The second is an error ε_l between the lengths of the two flexures. The parasitic twisting is then

$$\theta \approx \frac{q_1 \varepsilon_l}{bL} \tag{4.13}$$

Assuming that the ratio of displacement to length is small (it must always be less than unity) then Equation 4.13 will predominate. Clearly the paired lengths of the mechanism should be kept as similar as possible to minimize parasitic errors. This is done in fabricated devices by grinding the clamping faces of both

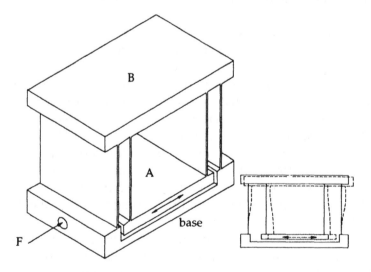

Figure 4.4 A compound linear spring; schematic representation of the mode of distortion (lower right)

the platform and base as a pair and setting the vertical lengths of the flexures using slip gauges before they are clamped. The errors are not always a serious problem in practice. Typically b and L will be about equal and both q_1 and ε rarely exceed about 1% of L, so errors of below 10^{-4} radians would be expected. It has also been found that variations in flexure thickness cause considerable errors and that these can be reduced by using rolled sheet from the same batch.

Compensation for the curvilinear motion can be achieved by combining two simple springs in a more symmetrical design to produce the compound spring illustrated by Figure 4.4. A simple linear spring mechanism is attached to the underside of a similar one. If both sets of springs are of the same material and dimensions, then application of a drive force to the primary platform A, which will be transmitted through both flexures, will result in similar deflections of all the leaf springs. Consequently, any change in height of platform A relative to its support B will be compensated by an equal and opposite change in height of B relative to the base. In theory this should result in a perfect rectilinear motion of A relative to the base. Because the springs are connected in series, equivalent to the addition of compliances, these devices will have a stiffness half that of the simple springs from which they are made. Their stiffness is given directly by Equation 4.10. Again such mechanisms have been assessed by Jones, 1956. It was found that, with spring lengths of 50 mm, displacements of up to 10 mm could be achieved with parasitic twists of less than 1 arcminute in all axes and a parasitic height variation of approximately 1 μm. This rectilinearity of better than 10,000:1 can be considered typical for small displacements. Variations on the general theme of compound design will be discussed in Section 4.3.4.

The deflection range of a flexure stage will be governed either by its stiffness, through a maximum force that can be applied to it, or by its elastic properties. Using the simple beam theory of Equations 2.12 and 2.18, this limit will occur when a maximum allowable tensile stress, σ_{max}, is reached at the section of maximum bending moment. Typically the allowable stress might be 0.3 to 0.1 of the effective yield stress for metal springs. For the simple spring mechanism, the drive load is shared by two springs so the maximum moment in each will be $FL/4$ if the drive is applied in the line of the mid-point of the springs, as recommended. Expressing the force in terms of the maximum stress and using Equation 4.10 then shows that the permissible displacement, q_{max}, is

$$q_{max} = \frac{\sigma_{max} L^2}{3Ed} \tag{4.14}$$

If the drive is applied directly to the platform, it might appear at first sight that the maximum bending moment is doubled. This is not the case because the legs constrain each other to prevent end rotations: a tensile force builds up in one, and a corresponding compressive force in the other to provide a moment that bends the springs back into the symmetrical S-shape. This force pair reduces the maximum bending moment in the springs to $FL/4$, as before. However, the situation is less favourable since the tensile stress in one leg must be subtracted from σ_{max} before calculating the available displacement. The tensile force has magnitude $FL/4d$, where d is the leg spacing, and the stress depends on the cross-sectional area of the springs. Thus we find a further reason for using a central drive. Since the compound spring consists of two simple springs in series, its maximum stress is equal in each section and governed by half of its total displacement. Hence the maximum permissible displacement is twice that calculated for one of its constituent sections by Equation 4.14. Again, symmetrical drives will lead to minimum stresses.

Consideration of the deflected shape of the leaf springs reveals that most of the bending occurs near the roots, with the middle sections remaining relatively straight. Thus, for only a small increase in drive direction stiffness, the buckling strength of the springs can be improved and errors due to their imperfections simultaneously reduced by clamping thick reinforcing plates to the central section of the springs. Again, Jones, 1956, provides further information. We shall not analyse this form but consider instead its logical extension, the notch hinge mechanism, in the next section.

The fabrication of leaf spring flexure devices introduces the problem of fixing the component parts. We have found with some flexures that there can be a discrepancy of 30% or more between the theoretical and actual stiffness, Smith, 1990. This is almost entirely due to the compliance of the screws and fixing plates that clamp the ends of the springs. The clamps will always cause some uncertainty both in the quality with which they simulate a truly built-in beam end and in the effective length of the beam. One way of overcoming these difficulties is to use a monolithic structure. However, it requires close control, and so is expensive, to machine good quality leaf springs from the solid.

4.3.2   Notch type spring mechanisms

Two of the desirable design features identified in the previous section are monolithic construction and the reinforcement of the central region of leaf springs. The benefits of both can readily be obtained by machining the flexure from a solid using notch hinges, see Figure 4.5. Two holes are drilled close together in a solid blank and then excess material removed to reveal the desired shape. This type of manufacture is ideally suited to CNC milling and boring machines. Another advantage is that the accuracy of the spring flexure is primarily dependent upon the accuracy of the centre of the holes with the material removed from the rest of the original blank being of little influence, so manufacturing costs can be quite low. Figure 4.6 shows a typical simple linear

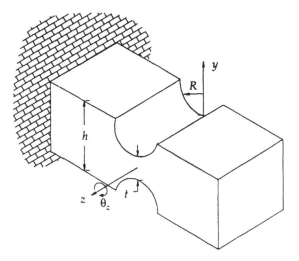

Figure 4.5 The notch hinge

spring mechanism constructed in this way. Assuming that each hinge distorts as if it were a perfect pin joint about an axis coincident with the thinnest part of the notch and that the thicker sections of the legs remain essentially undistorted, since $q_1 \approx L\theta_z$ the height variation is simply

$$h = L(1 - \cos(\theta_z))$$
$$\approx L(1 - \cos(q_1/L)) \approx \frac{q_1^2}{2L} \qquad (4.15)$$

Each notch acts as an elastic rotary bearing. This topic is discussed generally in Section 4.4, but it is appropriate to examine this particular case here. The hinge rotation, θ_z, as a function of applied bending moment, M, has been derived by Paros and Weisbord, 1965. For small deflections and assuming that the ratio $h/(2R+t)$ is near to unity, that is the notches are nearly semicircular,

$$\theta_z = \frac{9\pi R^{1/2} M}{2Ebt^{5/2}} \qquad (4.16)$$

Alternatively, for $t < R < 5t$, an approximation derived empirically from finite element studies, Smith *et al.*, 1988, gives

$$\theta_z \approx \frac{2KRM}{EI} = \frac{24KRM}{Ebt^3} \qquad (4.17)$$

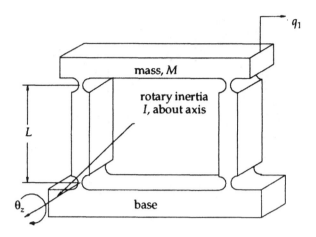

Figure 4.6 The simple notch type linear spring

where I is the second moment of area of the rectangular section at the thinnest part of the notch and K is a correction factor for the notch curvature modelled by

$$K = 0.565 \frac{t}{R} + 0.166 \tag{4.18}$$

Adopting the same arguments as used to derive Equation 4.14, the maximum moment that the notch can be permitted to support is

$$M_{max} = \frac{bt^2}{6K_t} \sigma_{max} \tag{4.19}$$

where

$$K_t = \frac{2.7t + 5.4R}{8R + t} + 0.325 \tag{4.20}$$

is the stress concentration factor caused by the circular notch shape. Then, combining Equations 4.17 to 4.20, the maximum allowable rotation of the hinge will be

$$\theta_{max} = \frac{4K}{K_t} \frac{R}{Et} \sigma_{max} \tag{4.21}$$

A simple linear spring has four notches. If a drive force F is applied in the line of the mid-point of the legs, the moment at each hinge is simply $FL/4$ and the displacement $\theta_z L$, so, for the Paros and Weisbord model, Equation 4.16, its stiffness, λ, in the drive axis is given by

$$\lambda = \frac{F}{q_1} \approx \frac{8Ebt^{5/2}}{9\pi L^2 R^{1/2}} \tag{4.22}$$

For the bending model of Equations 4.17 and 4.18, the equivalent form is

$$\lambda = \frac{F}{q_1} = \frac{2EI}{KRL^2} = \frac{Ebt^3}{6KRL^2} \tag{4.23}$$

Applying the drive force at the platform causes exactly the same behaviour pattern as with leaf springs: the moments at each hinge, and so the overall stiffness, are essentially unchanged, but the tensile stress in the hinge section somewhat restricts the bending that may be permitted. Note that because the bending is highly localized, good estimates of the overall stiffness of notch flexure mechanisms can be obtained by work-energy methods. Each joint stores strain energy proportional to its rotation squared which is readily calculated from Equation 4.16 or 4.17. The work that is injected is simply the product of average drive force and deflection, which is proportional to joint rotation. Equating these often yields the required relationships with little effort.

The maximum deflection of a notch hinge mechanism is normally governed by the peak stress in the thinnest section of the hinges. Using Equation 4.21 it will be

$$q_{max} = \theta_{max} L = \frac{4K}{K_t} \frac{R}{Et} \sigma_{max} L \tag{4.24}$$

for a centrally driven mechanism. As discussed above, driving at other points causes extra internal forces and reduces the deflection at which a peak tensile stress σ_{max} occurs.

With the leaf spring type flexures the load bearing capability is limited by Euler buckling. Notch type designs will undergo a change in stiffness due to compressive loading in the vertical direction, refer to Figure 4.6, given by

$$\lambda^* = \lambda - \frac{F_L}{L} \tag{4.25}$$

where F_L is the vertical load which is positive in compression.

If the ratio of load force to distance between the notches is greater than the drive direction stiffness then the effective stiffness is negative. This is the point at which catastrophic failure will occur upon the slightest deflection. Loading in the opposite direction will result in an increase in stiffness in which case it may become difficult to drive the platform.

General design ideas developed for the previous flexures such as splitting the flexure and separating the parts to increase the torsional stiffness, driving at a position midway between the notches and combining simple springs into a compound device to compensate for errors of rectilinearity apply equally to monolithic devices. As with leaf springs, the compound design has half the stiffness in the drive direction of the simple stage and, therefore, twice its deflection for a given maximum stress.

4.3.3   Kinematic analysis of linear spring mechanisms

The use of kinematic analysis and the idea of degrees of freedom or mobility to aid understanding of mechanism behaviour was introduced in Chapter 3. The same approach is also informative with flexure systems. As well as defining the number of coordinates required to describe the motion of the mechanism the mobility also gives the number of major resonances in a flexure design. For a plane mechanism, the mobility, M, is calculated from Grubler's equation, Equation 3.5 repeated here

$$M = 3(n-1) - 2j_1 \qquad (4.26)$$

where n is the total number of links in the mechanism including the base and j_1 the number of one degree of freedom joints having parallel axes of rotation. In the analysis of flexure mechanisms we consider a notch hinge to be a single degree of freedom pin joint. For leaf type flexures, each leaf spring is considered as a rigid link connected at each end by a single degree of freedom pin joint. Links with two degrees of freedom will be discussed later in this section. The elastic mechanism is regarded as equivalent to a kinematic chain with the same geometry together with a suitable number of springs that bias it to its equilibrium position.

Before analysing various mechanisms it is prudent to reiterate the discussion from Chapter 3. The mobility of a mechanism does not guarantee its behaviour, but is merely a guide. A designer can learn much about the real behaviour of a proposed mechanism by explaining why its mobility computes as it does. Elastic mechanisms are just as prone to surprising results as other types, but there are additional violations that we may exploit. For example, in classical design overconstraint should be avoided because it prevents strain-free assembly and so degrades performance. In monolithic design, we may readily

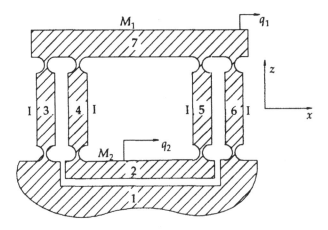

Figure 4.7   The notch type compound linear spring

construct strain-free overconstrained mechanisms, since all the members are, by definition the correct length to fit. Some overconstrained forms may have use as mechanisms. A flexure stage with several parallel legs may still deflect satisfactorily, with the extra legs both reducing the effects of manufacturing errors through elastic averaging and better distributing the transverse stiffness.

The simple notch spring mechanism shown in Figure 4.6 has a mobility 1, as expected for the elastic equivalent of a four-bar chain. The compound linear spring shown in Figure 4.7 has mobility 2, even though it is in practice treated as a compensated version of the simple spring. This happens because the coordinates of the primary and secondary platforms are actually independent. It is possible, for example, to set the primary platform in a fixed position and still move the secondary one to form a differential mechanism. The compound design operates satisfactorily as a single linear drive only because the springs automatically share the load and so couple the two degrees of freedom. The true pin-jointed kinematic chain could not be used in the same way. The double compound rectilinear spring of Figure 4.8 is simply a mirror image of Figure 4.7 about the primary platform axis. However, in this case there is only one degree of freedom and so the primary platform is properly constrained to move in a fixed path. From this analysis we expect superior performance from the double compound mechanism. As mentioned, this analysis can be interpreted to leaf spring mechanisms for which the superior performance of the compound and, particularly, double compound mechanisms has been established. A double compound leaf spring stage is shown in Figure 4.9.

Analysis of the compound rectilinear spring implies that more consistent behaviour may be achieved by the removal of one degree of freedom. This can

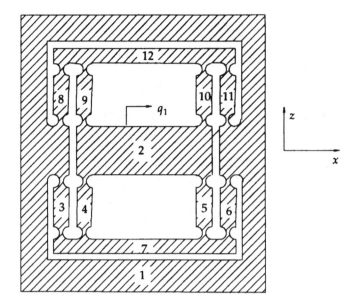

Figure 4.8 The notch type double compound rectilinear spring. Kinematic analysis indicates a single degree of freedom.

be achieved by coupling the motions of the two platforms. There are many ways of achieving this, Jones, 1951, Smith, 1988, one of which is shown in Figure 4.10. This mechanism contains the same rectilinear mechanism as Figure 4.7, represented by the links 1 to 7, with the addition a lever mechanism inserted between the primary and secondary platforms. With this design there is a kinematic, rather than merely elastic, coupling between the platforms. Any motion of the primary platform will impart a fixed motion onto the secondary and vice versa and so the degrees of freedom has been reduced by one. Inserting the values of 10 links and 13 joints into Grubler's Equation confirms this by indicating a mobility of one. Further, it can be seen that the ratio of the displacement of the primary platform, 2, to the secondary platform, 7, relative to the base, 1, can be approximated by the ratio of the dimensions b/a. As mentioned in Section 4.3.1, a perfect rectilinear motion will result if the distortion of the flexure springs is identical. This is achieved when the ratio b/a is equal to two and is known as 'slaving'.

So far only two dimensional systems with one defined freedom have been considered. Elastic mechanisms can be easily extended to two and three dimensional action. A common technique is to attach one simple linear spring perpendicular to the platform of another one, see Figure 4.11. Generally, more compact arrangements are achieved than when conventional sliding linear

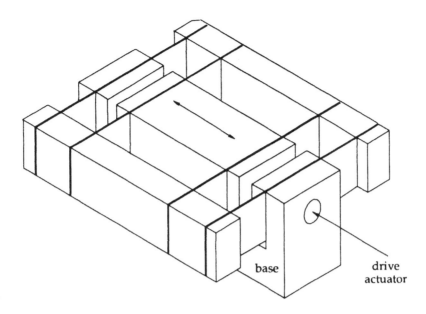

Figure 4.9   The leaf type double compound rectilinear spring

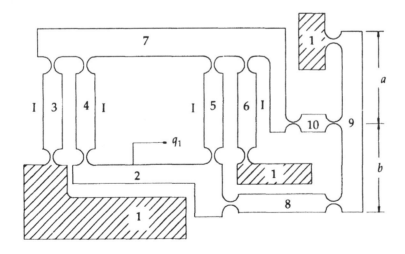

Figure 4.10 A slaved compound rectilinear spring. Shaded areas are rigidly fixed to the frame constituting a single link.

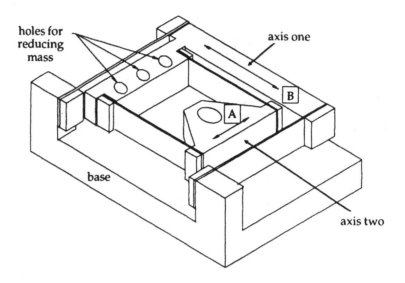

Figure 4.11 A two axis leaf type simple linear spring

motions are stacked. Alternatively, two degree of freedom hinges and flexures could be used. The simplest device is a square or circular cantilever beam. However, it does not have a stationary drive axis. Other elastic joints are shown in Figure 4.12. Figure 4.12a is simply two notch hinges with axes perpendicular (or at any desired angle) and Figure 4.12b is a universal circular flexure. The angular deflection of the latter due to an applied moment, M, can be approximated, Paros and Weisbord, 1965, as

$$\theta \approx \frac{20MR^{\frac{1}{2}}}{Et^{\frac{7}{2}}} \tag{4.27}$$

where R is the radius of the notch and t the diameter of the thinnest section. Such flexures can be used to form simple or compound linear springs to provide motion in two dimensions with stiffness

$$\lambda = \frac{F_x}{x} = \frac{2Et^{\frac{7}{2}}}{5L^2R^{\frac{1}{2}}} \qquad \text{(simple spring)}$$

$$= \frac{Et^{\frac{7}{2}}}{5L^2R^{\frac{1}{2}}} \qquad \text{(compound spring)} \tag{4.28}$$

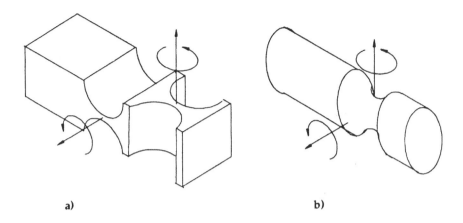

a)   b)

Figure 4.12 Two degree of freedom joints; a) a two axis notch type joint, b) a universal joint with coincident axes

Examples of basic designs are shown in Figure 4.13 and Figure 4.14. The first is a simple two axis spring. In principle its mobility can be determined from Kutzbach's formula, Equation 3.4. It has six members and eight identical joints. We plan that each joint has two freedoms, so the mobility of the system is –2. It is overconstrained and a simpler, more effective form ought to be possible. However, the notches will be relatively weak in torsion, so perhaps each joint has three freedoms. The mobility would then be +6. The extra freedoms partly reflect that each leg may now rotate about its own axis without influencing the platform. Perhaps, then, in our context it is better to consider only one joint on each leg to have the torsional freedom, so there are four joints with three freedoms and four with two freedoms. The mobility then calculates as +2, which is the answer we hoped for. Clearly, we may play dangerous games when analysing mobility, which is why it should be used not to confirm how a mechanism will behave, but as a means of posing questions whose answers will require careful study of how it might really behave. Here we merely observe that the torsional stiffness may well be quite a lot bigger than the flexural stiffness and that a real device will allow some freedom for the platform to rotate in its own plane. None of the three analyses suggested actually address this. We could perform exactly the same set of tests on a system with only three legs. Allowing torsional freedoms at all joints then gives a mobility of +6, which allowing for the axial rotation of each leg, provides three freedoms for the platform and perhaps better reflects real behaviour. Although the three-legged system appears to have advantages from this point of view, in practice the four-legged one is likely to be better because it is more symmetrical with respect to orthogonal drive directions. Our conclusion is that the design of flexures

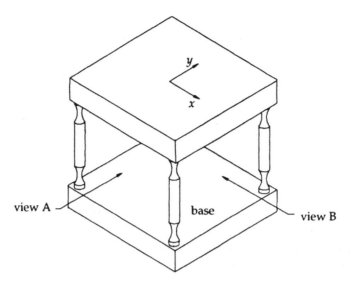

Figure 4.13 A two degree of freedom simple linear spring mechanism

having pivots with two degrees of freedom requires intuition and trust in the general concepts of symmetry, balance and superposition. One approach that lacks rigour but tends to produce viable designs is to look at the structure as a combination of two dimensional problems from particular views. For example, view A in Figure 4.13 will appear as a simple linear spring having 1 degree of freedom, as will view B. In both these views the system looks like a simple linear mechanism with notch hinges having twice the stiffness of the actual single joints. This provides a reasonable way of setting flexural stiffnesses, although it completely ignores any possibility of platform rotation. Note that the universal flexures shown in the diagram can be replaced by double notch hinges for more stable designs or the complete support flexure can be simply a rhombohedral or circular cantilever.

Figure 4.14 shows a compound equivalent of Figure 4.13. Both have the advantage over the normal piggyback arrangement of two single-axis translations that stiffnesses can readily be made equal in both axes. Parasitic errors should then be similar with both drives. A disadvantage of these more complex devices is the increasing difficulty of ensuring that the drive in one axis does not influence motion in the other. If both of the drives originate from the base, as they must in the design illustrated, then the x axis drive should have near zero stiffness in the y direction and vice versa. For small displacements this is usually approximated by placing a wobble pin between the actuator and platform, as discussed in Chapter 5. Other actuators having very high compliances in selected axes will be discussed in Chapter 7. The two axis

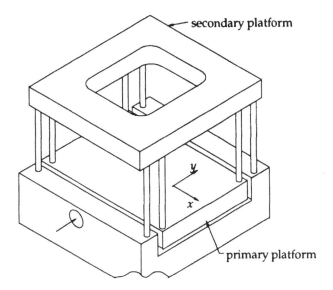

Figure 4.14 The two degree of freedom linear spring

flexure of Figure 4.11 has the apparent advantage that the drive for platform A can be mounted directly onto platform B, thus eliminating the need for a low compliance interface. However, the drive actuator places extra weight on the flexure and may introduce adverse non-symmetrical loads and undesirable heat sources. Again, the best compromise must be resolved in each case by the designer.

4.3.4 Additional considerations with compound spring designs

The symmetry of compound spring designs means that they have low susceptibility to positional errors caused by thermal expansion. Consider the mechanism of Figure 4.7. If the structure undergoes a uniform change in temperature, then, assuming that the material is homogeneous, a straight line passing through the axis of the primary platform and into the base will remain a straight line with only a change in length. Thus rectilinearity will be preserved at this point. Expansion of the legs 3 & 6 will cause the primary platform to be raised in the positive z direction while the expansion of legs 4 & 5 will move it an equal amount in other direction resulting in a zero net displacement, provided the support legs have identical lengths. A uniform temperature gradient along the z axis will also cause an identical thermal expansion in the support legs 3, 4, 5 & 6. A uniform gradient along the y axis will cause the flexures to bend about the x axis. The bending of legs 3 & 6 will cause the secondary platform 7 to both move out of plane and to rotate slightly, but again an equal bending of legs 4 & 5, results in virtually zero net rotation of platform 2 with respect to the base. A uniform gradient along the x axis will cause slight bending of the legs about the y axis, but the dominant effect will be that the legs will expand to different lengths. The length changes will cause the platforms to rotate about the y axis, but, since both the relative change and the angle shift it causes depend directly upon the leg spacing, platform 2 remains closely parallel to the base. Symmetry is broken by the different leg lengths so the kinematic path will have changed and the loads will no longer be equally shared by all the hinges. Nevertheless, the compound design is less susceptible even to this thermal error than is the simple one.

The more symmetrical double compound spring of Figure 4.8 is equally insensitive to uniform temperature changes. Its behaviour under uniform gradients in the x and y axes is similar to that of the compound form, except that the higher symmetry should ensure that the compensating distortions are better matched. It is likely to give better performance in practice. The presence of a uniform thermal gradient in the z axis is more problematic. An overall gradient will cause one set of the legs, 3 to 6 or 8 to 11, to be longer than the other, so the structure will consist of a combination of two slightly different compound springs. This imbalance may cause some increase in susceptibility to parasitic forces from the drive. If the gradient is uniformly decaying from the platform, as it might approximately be if there was a heat source on it, then symmetry is maintained and performance unaffected. Of course, with all the mechanisms increasing temperature tends to slightly reduce stiffness, both because the legs grow longer and because of thermoelastic effects. Still, these compounded forms are the obvious mechanisms for ultra-fine positioning or manipulation in a harsh thermal environment.

There is no clear-cut answer to the question of the best point at which to place a load on the platform. It should be at a position of as much symmetry as practicable, but it could be oriented so that its weight acts either in the z or y direction in Figure 4.7 and Figure 4.8. The former keeps the loading in the plane of the mechanism, which seems advantageous. However, some legs are then in tension and others in compression and there will be an additional moment proportional to the product of load and x axis deflection. This will reduce the symmetry of the joint stiffnesses and may therefore increase vulnerability to both parasitic forces and thermal disturbances and ultimately may result in the collapse of the springs as the stiffness is reduced to zero, see Equation 4.25. Loading along the y axis maintains very similar stresses at all the hinges. Unfortunately the bending of the leg pairs now add rather than compensate as happens with thermal gradients. Some out of plane deflection will occur, but if deep hinges can be used that may be tolerable in order to produce the highest symmetry for really precise, lightly loaded systems. The platform of a compound mechanism will always show some rotation about the x axis as it deflects under transverse load. The ideal double compound form will deflect without rotation.

Equation 4.22 shows that the stiffness of a simple flexure is strongly dependent upon both the thickness of the hinge and the leg length. Differentiating it using the chain rule and dividing by the intended stiffness gives an expression for the worst case fractional error

$$\left| \frac{\delta \lambda}{\lambda} \right| = \left| \frac{5}{2} \frac{\delta t}{t} \right| + \left| \frac{2 \delta L}{L} \right| + \left| \frac{\delta R}{2R} \right| + \left| \frac{\delta b}{b} \right| + \left| \frac{\delta E}{E} \right| \qquad (4.29)$$

The values for the error in both R and b are governed by any variations in the size of hole cut by the drill and the thickness of the original blank. Both are likely to be small in relation to the size of hole and blank respectively and can usually be ignored. Deviations in both the length between notches and in their thickness are equal to the positional error of the machining. Even with relatively precise machine tools, this has a rather alarming effect with thin notches. However, if the errors are random (for instance, when the holes are first centre drilled and there are no systematic errors in the machine tool) then the overall stiffness will be averaged over the number of notches and its variation will be reduced by the square root of the number of notches. The double compound spring with 16 notch hinges will clearly be the most predictable mechanism in terms of the effects of random machining errors. Deliberate overconstraint in monolithic devices could be introduced through extra legs and joints to increase the averaging. There is little evidence that it is worthwhile in practice.

4.4 Angular motion flexures

There are many instances when angular rotation over a small range is required. For angles of less than about $5°$, elastic design techniques may provide very effective solutions. Again, the simplest method is to use a short cantilever, Eastman, 1937. The maximum angle of rotation is limited by the highest stress allowable at the surface. In simple cantilevers, greater range is achieved by making a longer and proportionately wider flexure. The angular stiffness of the cantilever flexure of Figure 4.15 which may carry an axial load, P, is given by, Eastman, 1937,

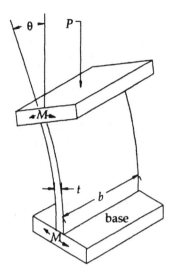

Figure 4.15 The simple cantilever hinge

$$\lambda_\theta = \frac{M}{\theta} = \frac{EI}{L} \qquad (\text{load } P = 0)$$

$$= \sqrt{PEI} \, \cot(L\sqrt{P/EI}) \quad (P \text{ compressive})$$

$$= \sqrt{PEI} \, \coth(L\sqrt{P/EI}) \quad (P \text{ tensile})$$

(4.30)

The stiffness in the presence of a compressive load rapidly drops to zero for larger values for the length of spring and load. This reflects the susceptibility of slender beams to buckling. However, for tensile loads, the angular stiffness increases with increasing load and the rise is most rapid for short cantilevers. It is clearly necessary to optimize the length based on the magnitude *and* direction of any applied loads.

The allowable stress in the outer surface of the flexure, σ_{max}, limits the maximum moment, so for zero axial load the maximum deflection is

$$\theta_{max} = \frac{2L \, \sigma_{max}}{Et} \tag{4.31}$$

Simple cantilevers have limited use as pivots because the centre of rotation moves with the angle of rotation. This error is much smaller for the notch hinge analysed in the previous sections, which is therefore now more commonly used.

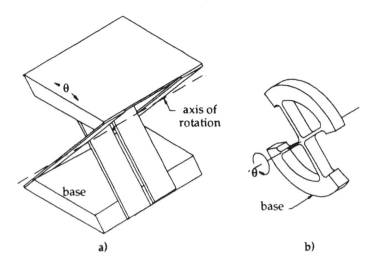

Figure 4.16 Angle spring hinges; a) the crossed strip hinge, b) the monolithic hinge

Another common spring pivot geometry is the cross strip flexure. Both fabricated and monolithic configurations are produced, usually of the forms shown in Figure 4.16. The angular stiffness of both forms in the absence of externally applied forces is given by Equation 4.30, taking the second moment of area as the total value across all flexures and L to be the 'diametral' length. A more detailed theoretical investigation of this type of flexure under a variety of load conditions is given by Haringx, 1949, which includes a formula for the shift of the axis with an increase in the angular displacement of the crossed strips. Monolithic flexures have much reduced levels of this parasitic error. An additional advantage is that the effective length of each spring for resistance to axial loading is halved giving a fourfold increase in capacity to withstand buckling along the axis and a twofold increase in resistance to bending failure, Young, 1989. The loadability is directly related to both length and width, and thus volume, for a given material. Also the load per unit volume can be increased by tapering the flexure towards the centre. If the flexures are in the form of a notch with the ends of the beams being twice the central thickness, t, the angular stiffness is, Haberland, 1978,

$$\lambda_\theta = \frac{1.67\,Ebt^3}{L} \qquad (4.32)$$

The increasingly complex shapes of these designs can be produced to close tolerance using electro-discharge and wire electro-discharge machining techniques under computer control. Ease of manufacture and the direct scaling

119

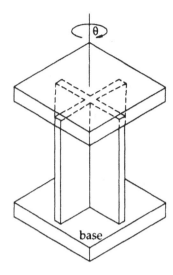

Figure 4.17 The cruciform angle spring hinge

of size with load capacity for a given material, has led to the commercial production of devices suitable for a wide range of applications.

Jones, 1955, has explored the twisting of beams about their longitudinal axis for use as angular spring hinges. A variety of cross-section shapes were studied, of which an equal sided cruciform section, see Figure 4.17, gave the best results. Compared to the assumption of a perfectly linear system with a fixed pivot point, typical off-axis errors expressed as a variation of angle were up to 50-75 μradian for a total angular deflection of 0.1 radian.

4.5 Dynamic characteristics of flexure mechanisms

The positioning of specimens or equipment to very high accuracies often involves feedback and sometimes requires the use of fine servo controls. Their speed of response will be limited by the natural frequencies of the flexure. The sensitivity of a mechanism to vibration also tends to be dependent on its natural frequencies. Flexure mechanisms are inherently spring-mass systems and so it is particularly important to characterize their general dynamic behaviour. A brief treatment is given here aimed at establishing how design parameters may affect natural frequency. A fuller examination of dynamic effects is given in Chapter 10.

For rotary mechanisms, with negligible frictional damping, the natural frequency is given simply by the square root of the ratio of total angular stiffness to total inertial mass (second moment of mass). Linear spring mechanisms contain elements that undergo combination of rotation and translation. Analysis is further complicated by the lack of simple kinematic descriptions of their motion. Even a simple four bar link movement is impossible to describe by a single-valued mathematical equation (it requires a transcendental function of a quadratic) and numerical methods must be used for the generalized case, see Tanaka, 1983. This happens because for most positions of a point on one link it is possible for the other two to occupy either of two positions (try it drawing it with a pair of compasses). For small movements of springs whose flexure supports are of equal length and initially parallel and perpendicular to the base, this simplifies to a linear and continuous problem. Consider again the simple notch type linear spring of Figure 4.6. The total mechanical energy of this system is the sum of the kinetic energy of the masses, T, and the potential energy stored in the springs, U, given by

$$
\left. \begin{array}{l}
T = \dfrac{M}{2}\dot{q}_1^2 + 2\dfrac{I}{2}\left(\dfrac{\dot{q}_1}{L}\right)^2 \\[1.5em]
U = \dfrac{1}{2}\lambda q_1^2
\end{array} \right\}
\tag{4.33}
$$

where I is the moment of inertia of the 'legs' about the base pivots and M is the mass of the platform. Defining the Lagrangian $\mathcal{L} = (T - U)$ and using the equation of generalized motion (see Section 2.2)

$$
\sum_{i=1}^{n}\left\{\dfrac{d}{dt}\left(\dfrac{\partial \mathcal{L}}{\partial \dot{q}_i}\right) - \dfrac{\partial \mathcal{L}}{\partial q_i} = Q_i\right\}
\tag{4.34}
$$

where n is the number of degrees of freedom, or mobility, and Q_i is the applied force in the direction of the coordinate q_i, yields for free vibration

$$
\left(M + \dfrac{2I}{L^2}\right)\ddot{q}_1 + \lambda q_1 = 0
\tag{4.35}
$$

This describes a simple harmonic motion having a natural frequency given by

$$
\omega_n = \left(\dfrac{\lambda}{M + \dfrac{2I}{L^2}}\right)^{\frac{1}{2}}
\tag{4.36}
$$

This equation can also be used as a close approximation to the natural frequency of a leaf type simple linear spring, using for I the value associated with a rigid body equal in dimensions and mass to the spring.

For the compound rectilinear spring, Figure 4.7, we require two coordinates to completely describe the motion of the mechanism, q_1 for the secondary platform relative to the base and q_2 for the primary platform relative to the secondary. Assume that all notches have the same stiffness and that all the legs are identical. The kinetic and potential energy of the system are then given by the equations

$$T = \frac{M_1}{2} \dot{q}_1^2 + \frac{M_2}{2} (\dot{q}_1 + \dot{q}_2)^2 + \left(\frac{M_L}{4} + \frac{I}{L^2}\right) \dot{q}_1^2 + M_L \left(\dot{q}_1 + \frac{\dot{q}_2}{2}\right)^2 + \frac{I}{L^2} \dot{q}_2^2 \qquad (4.37a)$$

$$U = \lambda (q_1^2 + q_2^2) \qquad (4.37b)$$

where M_L is the mass of one leg, I its second moment of mass about its centre of mass and λ is the stiffness of the primary platform relative to the base. Substituting these into Equations 4.34, the equations of motion for the compound spring are

$$\left(M_1 + M_2 + \frac{5M_L}{2} + \frac{2I}{L^2}\right) \ddot{q}_1 + (M_2 + M_L) \ddot{q}_2 + 2 \lambda q_1 = 0 \qquad (4.38a)$$

$$A\ddot{q}_1 + B\ddot{q}_2 + Cq_1 = 0 \qquad (4.38b)$$

$$(M_2 + M_L) \ddot{q}_1 + \left(M_2 + \frac{M_L}{2} + \frac{2I}{L^2}\right) \ddot{q}_2 + 2\lambda q_2 = 0 \qquad (4.39a)$$

$$B\ddot{q}_1 + D\ddot{q}_2 + Cq_2 = 0 \qquad (4.39b)$$

Since stable free oscillation can only be maintained with a single frequency and simple phase relationships between the subsystems, these can be solved simultaneously in terms of a simple harmonic oscillation by substituting a trial solution for q_1 and q_2 of the form (see Section 10.3.)

$$\left. \begin{array}{c} q_1(t) = A_1 e^{j\omega t} \\[2mm] q_2(t) = A_2 e^{j\omega t} \end{array} \right\} \qquad (4.40)$$

This leads to a pair of linear simultaneous equations from which the two natural frequencies ω_1 and ω_2 are given by the quadratic

$$(B^2 - AD) \omega^4 + C(A + D) \omega^2 - C^2 = 0 \tag{4.41}$$

It is not particularly helpful to solve this in general in terms of the inertias: it is better handled once numeric values for the parameters are known. The example serves to show how, even though Lagrangian analysis allows routine handling of the dynamic description, the control of resonances by adjusting inertias rapidly becomes beyond intuitive judgement. In many practical and successful mechanisms, the designers have had at best scant knowledge of the dynamic behaviour.

Finally, the equation of motion for the double compound rectilinear spring can be obtained by assuming that the deflection of the secondary platforms in this instance is constrained to be always exactly one half of the primary platform displacement relative to the base. If it is also assumed that the support legs between notches can be approximated as thin rods (I becomes $M_L L^2/12$), the equation of free motion of the primary platform is

$$\left(M_p + \frac{M_s}{2} + \frac{8M_L}{3} \right) \ddot{q}_1 + \lambda q_1 = 0 \tag{4.42}$$

where λ is the overall drive direction stiffness, M_L is the mass of each flexure, M_p the mass of the primary platform, I the inertia of each flexure and M_s the mass of each secondary platform. The single natural frequency is then given by

$$\omega_n = \left(\frac{\lambda}{M_p + \dfrac{M_s}{2} + \dfrac{8M_L}{3}} \right)^{1/2} \tag{4.43}$$

Again because it is a true single mobility system, the apparently more complex mechanism of the double compound spring is simpler to analyse than the compound rectilinear spring.

4.6 Case studies

4.6.1 X-ray interferometry

X-ray interferometry was first presented in 1965 by Bonse and Hart as a technique for the determination of crystalline perfection. Hart, 1968, pointed out that it could also be used as a measuring gauge having sub-Ångstrom resolution. Since then it has been used for the determination of the Avogadro constant, Deslattes, 1969, Deslattes *et al.*, 1976, the determination of lattice parameters in crystals, Becker *et al.*, 1982, linear and angular metrology, Becker *et al.*, 1988, Nakayama *et al.*, 1982; Alemanni *et al.*, 1986, and the calibration of high precision displacement transducers, Bowen *et al.*, 1990. Its relevance here is that it requires a short-range linear translation where, in one plane, parasitic twisting of 5 nanoradians is significant and there are tight constraints on the other parasitics. All published designs have used linear springs for the translation mechanism.

Conceptually the principle of x-ray interferometry is rather simple. An x-ray source is directed at an array of three blades of the same material which have accurately aligned crystallographic orientations, Figure 4.18. If the x-ray beam is directed onto the first blade at the correct angle (the Bragg angle) diffraction will occur and the x-ray beam will transmit through the structure. The first blade acts as a beam splitter and the second splits each of the beams again so that part of each converges. These interfere inside the third blade. The physics of the interactions of x-rays with crystals is far from simple, but the overall effect will be that if all the blades are properly aligned most of the power emerging from the third blade will be in the beam parallel to the input beam. If however a blade is moved slightly out of alignment then the forward diffracted signal will be attenuated and more power delivered in the other output beam. If one of the blades is moved parallel to the others with perfect rectilinear motion and the atomic planes of the crystal are not distorted in any way, then each displacement of one atomic lattice spacing causes the output signal to first reduce and then return to its original high intensity. Thus if the x-ray intensity

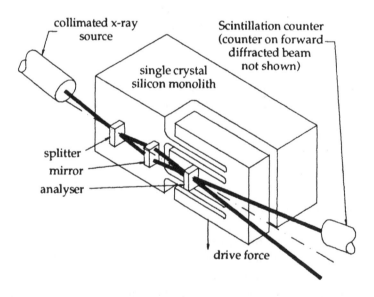

Figure 4.18 An x-ray interferometer manufactured from a single crystal silicon monolith

is monitored as one blade of the crystal is translated, a sinusiodal variation is observed. In this mode of operation the device behaves as an incremental grating with a pitch equal to the atomic lattice spacing. However, any twisting motion degrades the lattice alignment, causing a loss of contrast that makes variations in this intensity increasingly difficult to detect. The mechanical design challenge is thus to provide a linear translation of micrometre range that is easily controlled with sub-Ångstrom precision and has worst case parasitic rotations well below a microradian.

Single crystal silicon is the only convenient material for constructing the blades of a practical x-ray interferometer and the original designers used leaf spring flexures machined into the crystal to provide a monolithic device. Its stage was driven by applying a small force to the flexure either through a very weak spring or by using a small electromagnet. For small displacements, a simple or compound Jones spring was found to be adequate. However, for longer range motions Deslattes, 1969, chose to place one of the blades on the platform of a simple notch hinge linear spring made of brass. This approach demanded considerable skill in achieving the necessary alignment of the separated blades, whereas in the monolithic interferometer the blades are already near to alignment. Nevertheless it was possible to maintain usable contrast over a displacement range of 20 μm. The drive mechanism was simply a micrometer acting through a series of levers similar to the slaving mechanism presented in Section 4.3.2. Further to this Alemanni *et al.*, 1986, built a similar mechanism from the glass ceramic 'Zerodur' which has very high stability and an ultra-low coefficient of thermal expansion (see Chapter 8). The flexure hinges in this instance were elongated notches similar to those discussed by

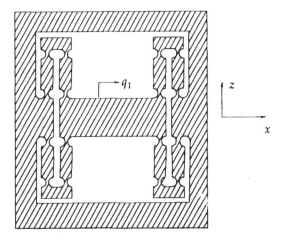

Figure 4.19 The linear translation mechanism of Becker *et al.*, 1987

Eastman, 1937. A piezo-electric actuator was used and a translation of 50 μm has been achieved.

More recently, Becker *et al.*, 1987, have presented an x-ray interferometer design that has a blade mounted onto a linear spring mechanism consisting of four simple springs in a symmetric arrangement similar to that shown in Figure 4.19. This strange mechanism was chosen to exploit its geometrical symmetries to obtain nice thermal characteristics and also so that its additional degrees of freedom could be used to align the crystal blades and to vary the axis of motion using electromagnetic pushers. The steel flexure was driven by a levered piezoelectric actuator and shown to be capable of operating with a range of up to 200 μm.

4.6.2 The measurement of friction between a diamond stylus and specimen

The frictional properties of small contacts under low load conditions are of considerable interest to instrument designers. A recent study, Liu *et al.*, 1992 1991, has monitored the frictional drag as lapped specimens of, for example, mild steel, aluminium and polished silicon were traversed by a stylus-based surface profilometer modified to allow continuous variation of the tracking force. A small magnet was attached directly above the stylus and surrounded by a solenoid coil. With correct positioning, the force on the magnet and so on the stylus is directly related to the energizing current in the coil. The main concern here is with the drag force measurement system, but other details are given as they are relevant to the discussions in Chapter 7.

An overview of the complete instrumentation is shown in Figure 4.20. The specimen is mounted onto the platform of a simple, notch hinge linear elastic translation which is deflected slightly by the drag of the stylus on the specimen. Connected to this is a modified Rank Taylor Hobson Talystep LVDI transducer with its measurement axis colinear to that of the spring (not shown in the figure). The Talystep uses leaf spring ligaments for its own motion, but they are much less stiff than the main platform. The latter was made of aluminium alloy with legs about 70 mm long and 5 mm deep and notches of radius 3 mm and central thickness 0.5 mm. The displacement sensor was calibrated to within 2% with a resolution of better than 0.5 nm. The platform stiffness was 1786 N m^{-1} giving a force resolution of better than 1 μN. The maximum normal force of the stylus was restricted to less than 5 mN during friction tests. The maximum friction coefficient of 0.28 encountered during experiments, then causes a worst case lateral deflection of the specimen of approximately 0.8 μm. Since the lateral deflection was always smaller than the stylus dimension, it was considered to have negligible effect in these tests. A paddle immersed in a bowl of oil was attached to the platform to provide damping.

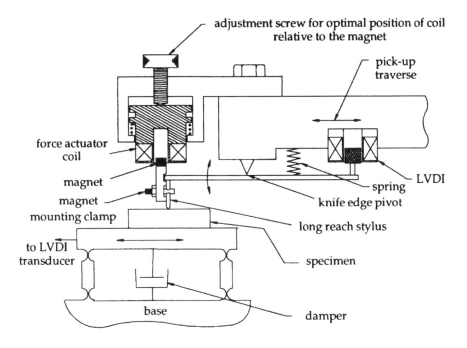

Figure 4.20 Schematic representation of stylus friction measurement instrumentation

An aluminium connector was clamped to the stylus shank of a Rank Taylor Hobson Talysurf 5 to support the magnet directly above the stylus and just above the stylus arm. A coil assembly was held over the magnet by a bracket clamped by set screws to the body of the pick-up. The coil was spring mounted to the bracket and held against an adjustment screw for setting the optimum magnet-coil position. This very simple construction is permissible because the force is independent of small displacements and so slight vibrations of the coil are not coupled into the measurement loop (for further discussion of the stylus force transducer see Chetwynd *et al.*, 1992).

To ascertain the value of the coefficient of friction, the stylus was traversed in the forward direction and a mean force calculated. Traversing the stylus in the reverse direction and noting the mean value then enabled calculation of the coefficient of friction as the difference of the two values divided by twice the normal load. This assumes the friction is independent of the traverse direction. The arcuate action of the stylus arm means that the dynamics are not identical in both directions of traverse but this is unlikely to affect friction to a significant extent.

The instrument proved to be sufficiently stable to allow convenient measurements of the frictional behaviour of a diamond stylus at loads in the region 0.5 to 5 mN. The overall precision obtained in practice for the measurement of drag forces parallel to the traverse axis was approximately 10 μN for a 10 mN range. Typical friction coefficients were found to vary from 0.07 for silicon to 0.25 for aluminium. These coefficients varied little with the load with a slight tendency to decrease with increasing traverse speed. Variations of the drag force about a mean value were observed to be dependent upon the surface finish. The results confirmed that a simple model relating the variation to the resolving of the normal force along the local surface slope was sufficient in this load regime.

References

Alemanni M., Mana G., Pedrotti G., Strona P.P. and Zosi G., 1986, On the construction of a zerodur translation device for x-ray interferometric scanning, *Metrologia*, **22**, 55-63.

Becker P., Seyfried P. and Siegert H., 1982, The lattice parameter of highly pure silicon single crystals, *Z. Phys. B. - Condensed Matter*, **48**, 17-21.

Becker P., Seyfried P. and Siegert H., 1987, Translation stage for a scanning x-ray optical interferometer, *Rev. Sci. Instrum.*, **58**, 207-211.

Bonse U. and Hart M., 1965, An x-ray interferometer, *Appl. Phys. Lett.*, **6**, 155-156.

Bowden F.P. and Tabor D., 1964, *The Friction and Lubrication of Solids*, Oxford University Press, chpt. 17.

Bowen D.K., Chetwynd D.G. and Schwarzenberger D.R., 1990, Sub-nanometre displacement calibration using x-ray interferometry, *Meas. Sci. Technol.*, **1**, 107-119.

Chetwynd D.G., Liu X. and Smith S.T., 1992, Signal fidelity and tracking force in stylus profilometry, *Int. J. Mach. Tools Manufact.*, **32**, 239-245

Deslattes R.D., 1969, Optical and x-ray interferometry of a silicon lattice spacing, *Appl. Phys. Letts.*, **15**, 386-388.

Deslattes R.D., Henins A., Schoonover R.M., Carroll C.L. and Bowman H.A., 1976, Avagadro constant – Corrections to an earlier report, *Phys. Review Letts.*, **36**, 898-900.

Eastman F.S., 1937, The design of flexure pivots, *J. Aero. Sci*, **5** (1) 16-21.

Haberland R., 1978, Technical advances through a novel gyro hinge design, *Symposium of Gyro Technology*, Deutsche Gesellschaft für Ortung und Navigation, Bochum, Sept, 18/19.

Haringx J.A., 1949, The cross spring pivot as a constructional element, *Appl. Sci. Res.*, A1, 313-332.

Hart M., 1968, An angstrom ruler, *Brit. J. Appl. Phys. (J. Phys. D)*, **1**, 1405-1408.

Jones R.V., 1951 Parallel and rectilinear spring movements, *J. Sci. Instrum.*, **28**, 38-41.

Jones R.V., 1955, Angle-spring hinges, *J. Sci. Instrum.*, **32**, 336-338.

Jones R.V., 1956, Some parasitic deflexions in parallel spring movements, *J. Sci. Instrum.*, **33**, 11-15.

Jones R.V., 1987, *Instruments and Experiencies*, Wiley & Sons., London

Liu X., Smith S.T. and Chetwynd D.G., 1992, Frictional forces between a diamond stylus and specimens at low load, *Wear*, **157**, 279-294.

Nakayama K., Tanaka M. Morimura M., 1982, Observation of x-ray interference from a two-crystal x-ray interferometer, *Bull. of NRLM*, **31**, 1-8.

Paros J.M. and Weisbord L., 1965, How to design flexure hinges, *Machine Design*, November 25, 151-156.

Plainevaux J.E., 1956a, Etude de déformations d'une lame suspension élastique, *Nuovo Cimento*, **4**, 922-928.

Plainevaux J.E., 1956b, Mouvement de tangage d'une suspension élémentaire sur lames élastiques, *Nuovo Cimento*, **4**, 1133-1141.

Smith S.T., 1988, Mechanical Systems In Nanometre Metrology, PhD Thesis, University of Warwick

Smith S.T., Chetwynd D.G. and Bowen D.K., 1988, The design and assessment of high precision monolithic translation mechanisms, *J. Phys. E: Sci. Instrum.*, **20**, 977-983.

Smith S.T. and Chetwynd D.G., 1990, An optimized magnet-coil force actuator and it application to precision elastic mechanisms, *Proc. Instn. Mech. Engrs.*, **204**, 243-253.

Tanaka M., 1983, The dynamic properties of a monolithic mechanism with notch flexure hinges for precision control of orientation and position, *Japan J. of Appl. Phys.*, **22**, 193-200.

Young W.C., 1989, *Roark's Formulas for Stress and Strain*, 6th ed., McGraw-Hill, London, 679-681.

5 DRIVE COUPLINGS AND THE MECHANICS OF CONTACT

The generalized concepts of kinematic design use models based on the point contact of infinitely rigid spheres, cylinders or flats. Clearly this is not possible in practice as materials have a finite stiffness and will deform upon any real contact. This chapter begins with a discussion of the behaviour of real, smooth contacts and the subsequent implications for precision design. It goes on to consider the implications for drives and locations and so to the design of couplings for transmitting forces in a desired direction without inducing significant parasitic forces in other directions.

Introduction

The distribution of stresses between two contacting elastic spheres was initially expressed mathematically by Heinrich Hertz and published in 1882. Since then there has been much practical and theoretical work relating the induced stresses and strains under a wide variety of load conditions. Much of this has been collated by Johnson, 1987, who presents the subject of contact mechanics in a mathematically rigourous and concise manner suitable for readers wanting a deeper insight than given here. Our objective in this chapter is to study the underlying implications of contact phenomena and to evaluate their significance for specific applications. Therefore problems have been simplified in an effort to aid physical intuition, without, one hopes, introducing any major errors in the results. Consider, for example, the oscillating transverse loading of a sphere vertically loaded against a flat. The problem is, as yet, intractable unless a lot of constraints, some of which are unlikely in practice, are imposed upon the model. However, solutions using these conditions indicate a hysteresis in the position of the sphere. So, although the model is rather divorced from reality, it points to possible difficulties in relocating kinematic clamping mechanisms should any perturbing forces exist. Many researchers have indeed been exasperated by their inability to reposition specimens, with apparent wandering at the nanometre level, and so it is possible that the study of even such simplified models may help us towards improved designs.

Having obtained an idea of the forces involved in stationary, sliding, and rotating contacts, the use of general kinematic principles for the design of couplings can be further investigated. A coupling is considered to be a device that is to transmit a force or torque from one part of a system to another in the desired degree of freedom *only*. In connecting a motor to micrometer, for instance, an ideal coupling should transmit a torque in the micrometer axis, but not any torques or forces in the remaining five degrees of freedom. A variety of coupling designs will be presented along with their relative merits and shortcomings.

5.1   Surface contact under different load conditions

Using simple models, this section investigates the effects of contact in the sliding and rolling of elastic bodies.

5.1.1   Stationary and sliding contacts of a sphere and a flat

The simplest type of contact is that between a smooth, frictionless sphere on a flat horizontal surface maintained in contact by a stationary, vertical force, P. A cross-section through the contact region is shown in Figure 5.1.

Hertzian analysis reveals that the contact radius, a, and the mutual approach of distant points in the two materials, δ, are given by the equations

$$a = \left(\frac{3PR}{4E^*} \right)^{1/3} \tag{5.1a}$$

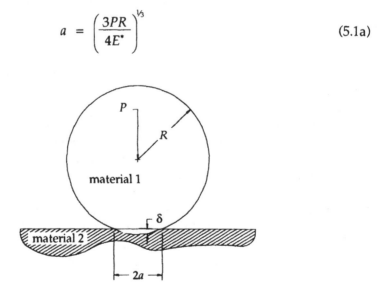

Figure 5.1   The contact of a sphere on a flat

$$\delta = \left(\frac{9P^2}{16RE^{*2}} \right)^{1/3} \tag{5.1b}$$

$$\frac{1}{E^*} = \frac{1 - v_1^2}{E_1} + \frac{1 - v_2^2}{E_2} \tag{5.1c}$$

where v and E represent respectively the Poisson's ratio and elastic modulus of the two contacting materials.

There will be a distortion with any finite load. The vertical stiffness, λ_v, of this interface is the inverse of the compliance defined as the derivative of deflection with respect to the applied load, that is

$$\lambda_v = \frac{dP}{d\delta} = \left(6E^{*2}PR \right)^{1/3} \tag{5.2}$$

This important equation shows that the interface stiffness depends upon the magnitude of the load, P. As a consequence, we may increase the stiffness of rolling element bearings, for example, through preloading of the balls by ensuring an interference fit between the rolling elements and the race surfaces. It is not therefore reasonable to assume that very light loads exist within ball bearings under conditions of small externally applied forces. This will be discussed in more detail in the next section. Additionally, it introduces a dilemma. Pure kinematic design relies on point contact, implying small contact forces and so low stiffness. Under these conditions the structure will be susceptible to vibration or other applied forces. However, using higher loads and larger balls to improve the stiffness will move us away from kinematic conditions, increase the frictional forces and may result in localized plastic distortion. From Equations 5.1 it is apparent that one compromise is to use materials with high values of elastic modulus, small radius balls and to reduce loads as much as possible. Spheres made from tungsten, sapphire and alumina and rolling element bearings containing silicon nitride balls are commercially available and suitable for this purpose.

Another sliding load condition commonly occurring with spherical contacts is that involving rotation about the vertical axis. In this case it is not possible to assume a friction free interface and the analysis becomes more complex. An analysis, based upon Hertzian contact, by Lubkin, 1957, shows that the torque about the vertical axis, M_z, is related to the coefficient of friction, μ, by

$$M_z = \frac{3\pi\mu Pa}{16} \tag{5.3}$$

The final load condition that we shall consider is that of the sliding of a sphere over a flat surface. In all kinematic analysis it is assumed that upon initial contact every part of the connecting bodies is free to move around in order to accommodate the errors in initial placement and manufacturing tolerances. In reality, this is only possible if the interface is friction free. As an example, consider the case of a sphere on a flat that is subject to both a vertical load, P, and a horizontal load, Q. This could represent the combined loads on a sphere as it is pushed into a vee groove or as it pulls across a flat in the process of locating into a triangular hole, see Section 3.1.2. A comprehensive analysis for varying loads has been carried out by Mindlin and Deresiewicz, 1953, where generalized techniques for computing relative displacement between obliquely loaded and unloaded spheres are outlined. This analysis reveals some interesting characteristics. Results indicate, rather obviously, that during vertical loading the contact area gradually increases, with the radius, a, being derived from Hertzian analysis. If no slip occurs, the surface will be under a radially symmetric traction. Upon application of a horizontal force, an infinite stress is predicted at the periphery of the contact region. This will be relieved by slip around the edge and, as the tractive load, Q, increases further, an annular region will widen towards the centre with the inner radius of slip, c, given by

$$c = a\left(1 - \frac{Q}{\mu P}\right)^{\!1/3}$$
(5.4)

At the point $\mu Q = P$, the inner radius reduces to a point and gross slip, or sliding, occurs. At this point the problem is indeterminate. Theoretical analysis of the stress distribution under a spherical indenter during gross slip was first presented by Hamilton and Goodman, 1966, and these results are simplified and expressed in explicit form by Hamilton, 1983. This will be discussed shortly. A complete analysis of the variety of possible load conditions is beyond the scope of this book. However, analysis of just the simplest loading and unloading scenario provides practical insight and leads to guidelines for ultra-precision practice. We shall consider the situation in which both the sphere and flat are made from the same material, that $\mu Q < P$ and that the vertical load has been applied first. It can then be shown that the relative tangential displacement between the sphere and flat, δ_t, is given by

$$\delta_t = \frac{3(2 - v)\,\mu P}{8Ga}\left(1 - \frac{c^2}{a^2}\right)$$
(5.5)

where the shear modulus G is for an isotropic material and that the tangential stiffness, λ_t, is

$$\lambda_t = \frac{4Ga}{2-\nu}\left(1-\frac{Q}{\mu P}\right)^{\frac{1}{3}} \tag{5.6}$$

If the normal force is maintained constant and the tangential force linearly reduced to zero, then there occurs a permanent set, δ_u, of magnitude

$$\delta_u = \frac{3(2-\nu)\,\mu P}{8Ga}\left|\frac{c^2}{a^2}-1\right| \tag{5.7}$$

As an example, consider a steel ball of 10 mm diameter and shear modulus of 80 GPa touching a flat surface with which it has a coefficient of friction of 0.33. A vertical load of 10 N is applied resulting in a contact radius of 69.3 μm. If there is then a temporary horizontal load of 2 N, giving an inner slip radius of 50.8 μm, Equation 5.7 predicts a permanent shift of approximately 180 nm! Associated with this permanent shift will be a residual stress that may introduce a temporal error due to material creep. This can also be calculated from equations given in Mindlin and Deresiewicz, 1953.

This example highlights some inconsistencies that may arise if pure kinematic design techniques are employed at the nanometre level. Changes in position due to frictional effects are unlikely to be as serious as indicated by the above analysis. This is because there are six contacts in complete kinematic constraint. If the specimen is positioned by dropping vertically into the grooves of a type II clamp and the centre of gravity (or other force closure) passes through the centroid of the three ball mounts, all of the distortions will be of equal magnitude and the lateral position of the specimen should be repeatable to better than the predicted error for an individual contact. However, the vertical position will be dependent not only upon the load but also upon its history. Although this is a toublesome error source which many designers acknowledge informally, it is rarely reported, although the problem is mentioned by Hocken and Justice, 1976. Various efforts have been made to quantify such effects, Johnson, 1955, Goodman and Brown, 1962.

In static Hertzian contact the peak shear stress lies centrally below the surface at a depth of about half the contact radius. Another important consequence of one sphere sliding over another is that the position of the peak stress rises towards the surface as the friction coefficient increases. It has also been shown that the point at which yield occurs moves towards the trailing edge of the moving sphere and that the maximum principle shear stress occurs at the surface when the friction coefficient exceeds 0.27, Hamilton, 1983. Even if yield stress is reached, there is usually little practical effect provided the position is surrounded by material that remains in its linear elastic region. Once the peak stress occurs at the surface this protective shell is no longer there to act as a reinforcement. Thus the likelihood of surface damage in the contact area rises rapidly as friction increases. It can be reduced by adding a boundary lubricant, where possible. This will usually reduce the coefficient of friction to below 0.2 and so greatly improve the life of the contact.

It is also worth noting that the ratio of *initial* tangential stiffness to the normal stiffness is given by

$$\frac{\lambda_t}{\lambda_v} = \frac{2(1-v)}{2-v} \tag{5.8}$$

This ratio is independent of the friction coefficient and ranges from unity for a Poisson's ratio of zero to 2/3 for $v = 0.5$

5.1.2 Rolling contacts

As one might intuitively expect, the resistance to pure rolling between two elastic solids is considerably less than that due to sliding. Theoretically, two similar materials in elastic contact should have zero rolling resistance because the strain energy stored as they compress together at the leading edge of the contact zone is returned as stress levels reduce towards the trailing edge. In practice there will be a finite resistance even at low loads due to loss mechanisms such as dislocation movement, viscoelasticity and creep in solids. In the absence of precise theoretical arguments, it is usual to derive an empirical loss factor, α, and the equation

$$\mu_r = \alpha\frac{a}{R} = \frac{M}{P} \tag{5.9}$$

where μ_r is the coefficient of rolling resistance, M is the torsional moment and P the normal load between the sphere and surface. Note that μ_r is expressed in units of length and, although closely associated with the coefficient of (sliding) friction, should not be considered a direct measure of friction. In practice the loss factor is rarely more than a few percent, giving a fairly typical coefficient of less than 10^{-4} m, see also the example discussed below. This value, it should be noted will also include added effects caused by surface roughness, Lubkin, 1951, Eldridge and Tabor, 1955, Tabor, 1955.

For a ball bearing rolling in an inner race which conforms to the shape of the ball, the relative velocity varies across the contact region. This can easily be visualized by imagining an extreme case where the race goes half way around the sphere. If the ball is rotated in this channel with an angular velocity ω and the bottom of the ball adheres perfectly at the lowest point, then the surface of the ball at the height R will be sliding over the ball race surface at a velocity ωR. This was first identified by Heathcote, 1921 and subsequently became known

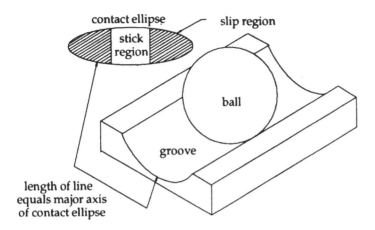

Figure 5.2 The contact of a ball rolling in a conforming groove

as Heathcote slip. There is a region at the root of the groove where no slip occurs with microslip regions extending either side of it out to the extremity of the contact region. Unlike the microslip discussed in the previous section, this does not form an annulus of slip, but linear regions discriminate the two regimes, see Figure 5.2.

An upper bound on the coefficient of rolling resistance occurring under conditions of high conformity and low interface friction, μ, is given by

$$\mu_r = 0.08\mu \frac{b^2}{R} \tag{5.10}$$

where b is the half length of the major axis of the elliptical contact region as observed from the vertical elevation. This could be up to the order of the radius, although usually less, from which, assuming an interface friction coefficient of approximately 0.2, rolling friction should not exceed 0.016 m. Reducing the contact to that subtending one radian, reduces the coefficient by a factor of around ten. We see that ball bearings can have a considerably higher coefficient of rolling friction than a simple ball rolling on a flat surface, although it is still much smaller than sliding contact friction.

The influence of surface finish and hysteresis has been ignored in all of the previous discussion. It is well known that poor surface finish degrades the performance of bearings and increases frictional influences. There does not appear to be any systematic, quantitative study of this. Intuitively, smooth surfaces are more desirable and should be used if possible. Some support for this view is provided by Halling, 1956, who found that the rolling resistance between a sphere and flat reduces after repeated cyclic loading. This is further discussed in Section 5.3.2. Tabor, 1955, also points out that although sliding may contribute to the wear of rolling element bearings, the rolling friction will be strongly influenced by hysteresis losses in the contacting materials and may be particularly significant for shallow groove bearings.

5.2   Linear drive couplings

Several different techniques for the rectilinear transmission of forces are presented in this section. The driving element, usually a micrometer or feedscrew arrangement, will not push reliably in a pure rectilinear path but will show some variation about a mean line. Additionally, it is unlikely that the direction of the drive force is colinear with the axis of the driven stage. Our experience with micrometer-drive stages has invariably revealed slight variations in the pitch of the spindle (pitch error) and a larger perturbation from rectilinearity that is periodic with its rotation (spindle error). The latter error encourages cross-axis translational or pitch or yaw motions. The former affects positional accuracy in the drive direction and induces varying forces through the impressed changes of speed. With standard commercial drives parasitic motions of up to $10\,\mu m$ are commonly encountered. Figure 5.3a shows a typical micrometer drive pushing a specimen carriage. Any parasitic errors will be directly transmitted through friction at the interface. The use of two flats may aid the transmission of large forces, but is not a kinematic coupling and, if there are parasitic spindle errors, the actual contact point may move during displacement of the carriage. Misalignment between the axes of the micrometer and the carriage will result in contact at the edge of the micrometer spindle. Subsequently, as the micrometer spindle rotates there will be frictional forces perpendicular to the drive axis which will induce additional parasitic errors. This can be reduced by making the end of the spindle spherical to ensure a 'point' contact, Figure 5.3b. Although eliminating the possibility of edge contact, it does not reduce the influence of spindle error. Also it introduces a new problem if the spindle is rotating as the drive axis must pass through the centre of the sphere. Failure to ensure this will result in the sphere describing a circular path with a radius equal to its eccentricity about the spindle axis. Since, to some degree, such errors are inevitable, it is desirable that they be decoupled from the driven stage. Kinematically, this can be expressed as the requirement for a single degree of freedom coupling with infinite stiffness in the axis of the *driven stage* and zero stiffness in the other five degrees of freedom. The simplest possible method, and quite often the best, is to separate the two by a wobble pin, see Figure 5.3c.

If the length of the wobble pin between the contact lines at each end is L then an eccentric rotation of the spindle, e, will result in the wobble pin describing a conical arc of half angle, e/L. Clearly the angular misalignment will be in

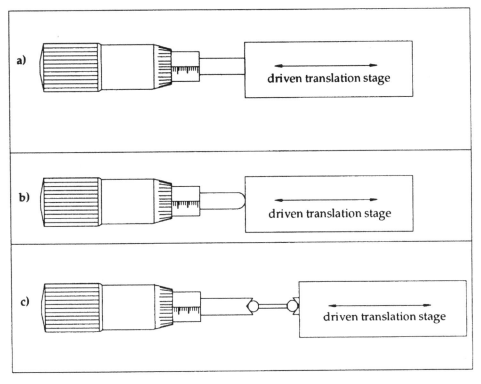

Figure 5.3   Simple drive couplings between a micrometre and linear motion slideway; a) unidirectional micrometer drive, b) ball nosed micrometer (non-rotating), c) wobble pin drive.

inverse proportion to the length of the wobble pin. The conical support cups will still transmit a relatively large couple in all rotary degrees of freedom. Ideally the contact areas should be reduced to points and the length of the wobble pin increased to infinity. A possible compromise is to use a ball between two flats. The stiffness of this coupling is approximately half that given by Equation 5.2. The main problem with this configuration is to keep the ball in position when there is no applied force. This may be overcome, providing that a slight dead-zone or backlash is tolerable, by holding it in place with a relatively weak spring as shown in Figure 5.4. The ball is held in a vee groove, through which it protrudes, by a plate that is pressed against the ball using a spring. When the drive contacts the ball, it will first have to overcome the spring force before the plate moves towards the rear of the coupling housing. Finally, provided that the distance between the sprung plate and housing is less than the protrusion of the ball from the vee groove *and* that the spring force is less than the frictional force required for motion of the carriage, the drive ball and supporting plate will form almost the ideal coupling configuration. Like all the

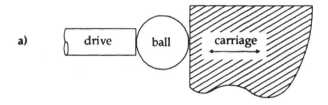

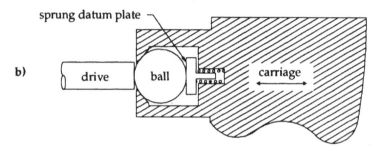

Figure 5.4 The ball and flats coupling; a) ideal ball coupling, b) practical design for retension of ball

designs discussed so far, it has the singular disadvantage that it cannot be used to drive the carriage in both directions. Consequently, it is necessary to use a spring, weight, or similar method, to provide a return mechanism.

An alternative geometry for kinematic coupling uses two crossed rollers. The contact mechanics of two rollers is the same as of a ball and flat with the radius of the contact region, mutual approach of distant points and stiffness given by Equations 5.1 and 5.2 respectively. If the two rollers are not free to rotate about their own axes, they will transmit frictional forces in the plane perpendicular to the drive axis, which contains the most important parasitic errors in most screw thread and hydraulic drives. The extra two degrees of freedom can be provided by mounting each roller in ball bearings. Using two parallel rollers mounted either side of the cross-axial one provides a bidirectional drive, Figure 5.5. The inclusion of four additional bearings in the force loop further reduces the stiffness of the coupling. A stiffer variant is found by inversion of this idea: three rods are rigidly held with the ball races mounted centrally. Contact occurs between the outer races of two perpendicular bearings. In theory, the stiffness should be twice that of the form shown in Figure 5.5. The cross-roller designs still suffer from backlash and may have a slightly higher friction coefficient in the y and x axes than the ball coupling because of the rolling element bearings. Usually it is reduced simply by lubricating the coupling and bearings with a thin instrument oil.

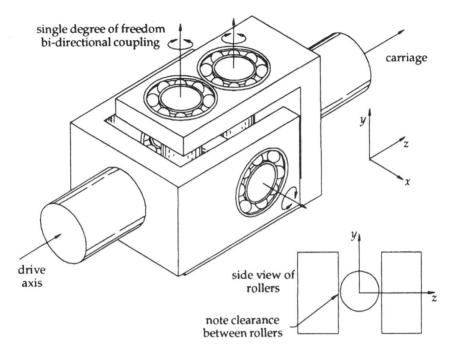

Figure 5.5 A bidirectional linear axis coupling

If a pair of crossed rollers are joined together with their axes in one plane, we obtain the familiar Hooke's joint. This eliminates backlash but loses three degrees of freedom, two linear in the x and y axes and one rotary about the z axis. The linear constraints can be relieved by using two joints separated by a long wobble pin, analogous to the drive shaft in a car.

If less stiff drives are acceptable or closed loop control is to be implemented, then there are a number of alternatives that may be of use for linear translations. A logical extension of the concept for a stiff drive is to think of a completely solid coupling. However, it is stiff in coordinates other than the one linear degree of freedom that is required. Stiffness in the other freedoms can be reduced by thinning the cross-section. Taken to its logical conclusion, this results in a wire solidly attached to both the drive and carriage. A wire buckles in compression and so it must be kept in tension, making the simplest wire drive unidirectional. Tensioning springs could be used to enable bidirectional operation, but it is often relatively easy, and more effective, to provide a push-pull wire drive. The stiffness of a wire along its axis is simply the product of its elastic modulus and cross-sectional area divided by the total length. Similarly the torsional rigidity is the product of rigidity modulus and polar moment of inertia divided again by the length of the wire. For a circular wire

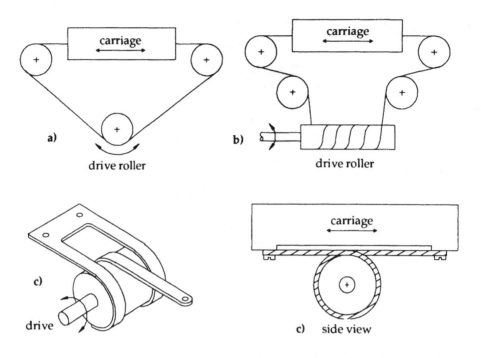

Figure 5.6 The wire and ribbon drives; a) perpendicular wire drive, b) parallel axis wire drive, c) the ribbon drive.

(polar second moment of area, $\pi r^4/2$) the ratio of the linear to torsional stiffness is inversely proportional to the square of its radius. Consequently, it is favourable to use a material having a high elastic modulus and to make the wire as thin as is practicable. The linear stiffness of the wire drive perpendicular to the drive axis is more difficult to calculate. For small displacements this can be approximated by the ratio of the tension to the length. Clearly, this will depend on the forces required to drive the carriage which are often of a variable nature. However, for applications where the loads are nominally constant such as ruling engines, this can provide an ideally smooth and continuous coupling, Stanley *et al.*, 1968. There are many variations about the wire drive of which two are shown schematically in Figure 5.6a & b.

Figure 5.6c shows a further variation on the wire drive in which a ribbon is used. Similar arguments apply for stiffnesses in all degrees of freedom except for a further constraint imposed on yawing motions for the carriage. This technique is commonly used to position the reading heads to sub-micrometre precision on computer magnetic disc drives.

5.3 Rotary drive couplings

The smooth transmission of rotary motion from motor drives to feedscrews or rotary tables is another common requirement in instrument design. Again the coupling can be specified as a single degree of freedom kinematic constraint. Theoretically it can be provided by single point contact applied perpendicular to a fixed rotary axis at a fixed radial distance. Couplings of this nature are very difficult to implement in practice because of the difficulty of avoiding parasitic torques in other axes. (A related device is the lever pivot for the generation of small angular deflections. This is discussed in Section 6.2.) The simplest practical solution is to use two-point contact or a continuous coupling having a compliance high enough that its effect in all other degrees of freedom can be neglected. The knife edge support is of this class and, for low loads, will have a very low friction coefficient. Using this simple device, very high precision balances and other force measurement instruments have been produced. An interesting early application of the knife edge pivot in a force measurement is that of Irving Langmuir's balance for the compression of oil films one molecule thick, Langmuir, 1917, which is discussed as a case study at the end of this chapter.

5.3.1 Elastic rotary couplings

Because there are difficulties maintaining the positional register of single contact rotary constraints, it is more common to use a specially designed continuous coupling. Typical forms for precision drives are the bellows, spiral and cruciform couplings. The principle of operation of a typical bellows coupling can be inferred from the name, an example is shown in Figure 5.7a. Mathematical analysis of its compliance in all degrees of freedom would be extremely complicated and probably of little use. However, a simplified approach does provide useful design guidelines. Figure 5.7b represents the cross-section of typical coupling which consists of an alternating series of thin-walled convex and concave half-toroids attached at their rims. The torsional stiffness of this arrangement about its axis, λ_θ, can be roughly approximated by that of a cylinder of the same wall thickness and of outside diameter, $2r_o$, equal to the average diameter of the bellows. The length of the cylinder is simply the number of hemispherical segments, n, multiplied by their cross-sectional length, πR. Reiterating Equation 2.7 for an isotropic material, the elastic modulus, E, is related to the rigidity or shear modulus, G, by

$$G = \frac{E}{2(1 + v)} \qquad (5.11)$$

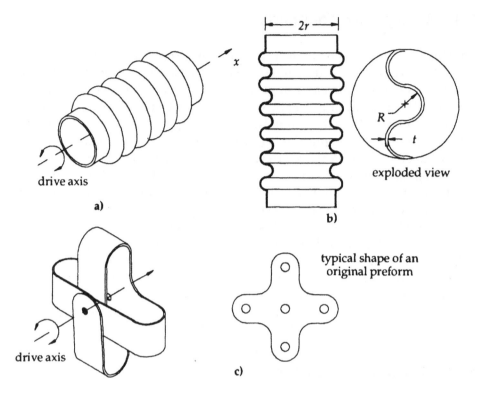

Figure 5.7 Simple elastic rotary couplings; a) The bellows, b) Cross section of bellows, c) Folded-cross or cruciform coupling.

Equation 4.9 may then be used to give

$$\lambda_\theta = \frac{E(r_o^4 - r_i^4)}{4(1 + v)nR} \tag{5.12}$$

Clark, 1950, shows that the stiffness of this type of bellows in the axial direction is given by

$$\lambda_x = \frac{Et^2}{0.577 \, Rn \, (1 - v^2)^{1/2}} \; ; \; R \geq 10t \tag{5.13}$$

where t is the thickness of the material, that is the difference between the inner and outer radii of the equivalent cylinder used in the previous analysis.

Equation 5.12 may be re-written in terms of the outside radius and the wall thickness (since $r_i = r_o - t$). Providing t is small compared to r_o, the resulting expression may be expanded as a power series and then truncated after the

144

term linear in t. For thin walled bellows the torsional stiffness is approximated by

$$\lambda_\theta \approx \frac{E r_o^3 t}{(1 + v)nR} \qquad (5.14)$$

Consequently, the ratio of the angular to axial stiffness is given by

$$\frac{\lambda_\theta}{\lambda_x} = \frac{0.577 r_o^3 (1 - v^2)^{1/2}}{t(1 + v)} \qquad (5.15)$$

Clearly we can maximize this ratio by choosing the average radius of bellows to be as large as is practicable and reducing the thickness to a minimum. Stiffness in the two other rotary axes can be analysed in a manner similar to that for the spring drive, below, in which it is shown that the stiffness is approximately proportional to the tension divided the length of the spring. A long coupling will tolerate significant extensions or contractions making it useful for connection to micrometer drives that require a change in length between thimble and motor/gearbox. Tension induced in the coupling by its change of length will, however, increase the degree of coupling to rotary parasitic errors from the drive.

The various types of bellows drives can only be produced by specialist manufacturers and are relatively costly. A cheaper alternative is to produce a folded-cross coupling as shown in Figure 5.7c. In this, a simple cross is produced from a piece of sheet metal, plastic or reinforced rubber and the ends bent over to produce the cruciform shaped coupling. The torsional stiffness is difficult to predict because the shear geometry is not simple. The interrupted circumferential section will reduce stiffness compared to a bellows, but this is countered by the smaller length which will increase it. Axial stiffness and bending stiffness will relate to the flexure of the arms treated as cantilevers from the shafts.

The very low axial stiffness of these couplings is desirable from a kinematic standpoint. However, in many practical designs the shaft from a motor drive has little support perpendicular to its axis. Under this circumstance, the shaft may sag significantly, often visibly. Consequently, a simple helical spring having a relatively high lateral stiffness is commonly used as a rotary coupling. For a spring of rectangular cross-section, as shown in Figure 5.8, the axial stiffness, λ_x, is given by the equation, Young, 1989,

$$\lambda_x = \frac{(8Gb^4)\,(a/b - 0.63(\tanh\,(\pi b/2a) + 0.004))}{3\pi R^3 n} \qquad (5.16)$$

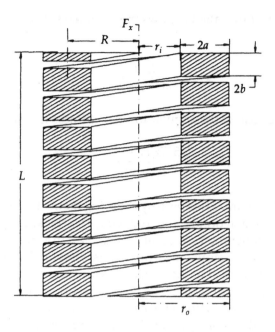

Figure 5.8   Cross section of a rectangular bar helical spring for use as a rotary coupling

For springs of small pitch angle and deflection, the torsional stiffness, λ_θ, of this device can be approximated by the equation, Wahl, 1963,

$$\lambda_\theta \;=\; \frac{4Eba^3}{3\pi n(r_o + r_i)} \tag{5.17}$$

For springs of square cross-section, the ratio of angular to linear stiffness is proportional to the square of the mean radius, favouring, again, spring couplings of large diameter. For a more detailed analysis of helical, and a few other, spring mechanisms the reader is referred to Wahl, 1963.

One variant of the helical spring coupling has a distinct identity of its own. The so-called spiral coupling is slightly misnamed for it is actually a helical form. It is made from a thick walled cylinder by making a series of helical saw cuts around its middle section. This section then behaves very much as a helical spring coupling, while the ends provide natural bosses for connecting it to drive shafts. Because it is a monolithic device it will generally provide better stability at a greater flexibility than a coupling fabricated from conventional springs. Figure 5.10 includes a view of a spiral coupling.

5.3.2 Rolling rotary couplings (the knife edge)

So far we have looked at rotary couplings that are essentially solid shapes of a geometry designed to reduce out of axis forces to insignificant levels. An alternative approach is that of rolling of a sphere over another sphere or a flat surface. For rolling contacts it is not possible to describe the angular orientation without also implying the position of the contact. Although the rolling ball appears to have five degrees of freedom, the frictional force required to prevent sliding imposes a second constraint. Classically any system in which the number of coordinates required for a complete description of the position of all its elements is greater than the degrees of freedom is known as non-holonomic. This condition prevents us using the usual rules of kinematic constraint and requires a more pragmatic approach. Consider, for example, a pair of spheres attached to each other by an arm of length L contacting a perfectly flat surface with a frictional traction sufficient to prevent sliding, as for the pivot of Figure 5.9. Application of a torque about the axis joining the centres of the two spheres will theoretically result in pure rolling. As briefly discussed in Section 5.1, the coefficient of rolling resistance, given as the resistance torque divided by the applied normal load, is normally less than 10^{-4} m. The questions that now arise are what radius of ball should be used and what constitutes an acceptable load. This is of great importance if the pivot is to be used for a balance or other similar precision instrument. The second question is relatively easily answered using Hertzian analysis. For ductile materials the maximum shear stress, τ_m, will indicate the closeness to plastic deformation and so to change in pivot behaviour. The maximum tensile stress, σ_m, would similarly be used for brittle materials. The shear component, τ_m, occurs below the centre of the contact at a depth of approximately half of the contact radius given in Equation 5.1. The maximum tensile stress occurs at the surface around the circumference of the contact circle. The magnitudes of these components can be calculated, using the notation of Section 5.1, from the equations

$$
\left.
\begin{aligned}
\tau_m &= 0.31\,P_o \\[2mm]
\sigma_m &= \tfrac{1}{3}(1 - 2v)P_o \\[2mm]
P_o &= \frac{3P}{2\pi a^2}
\end{aligned}
\right\}
\tag{5.18}
$$

The maximum stress should certainly not exceed the yield strength of the spheres and ought to be smaller than this value by a factor of at least three, and preferably ten.

Choice of ball radius is complicated by conflicting requirements. Firstly, from Equation 5.9 it is apparent that the lowest value of rolling friction is obtained by minimizing the ratio of contact radius to ball radius, which is satisfied by maximizing the radius of the ball. From considerations of stress concentrations in the contact zone, it is apparent from Equations 5.18 and 5.1 that, again, the ball radius should be large. However, as the ball radius increases, so too, does the translational displacement of its centre associated with a given angular displacement. The first two factors vary with the ball radius as a two thirds power law while the latter varies with unity power. It is therefore reasonable to reduce the ball dimension as much as possible, subject to the yield stress criteria. In all cases, frictional moments will be reduced by keeping loads as small as possible and using spheres made from very stiff materials, see Section 5.1. An interesting and important point to note is that Halling, 1959, found surface finish to have little effect on rolling after a very brief running-in period. It is, however, very important to provide for such running-in prior to the calibration of instruments.

Rolling resistance is *not* a direct indicator of the precision of a knife edge bearing, although it must be small for the bearing to be effective. The loads that provide the bearing torque may be small compared to those, from weights or retaining springs, holding the knife edge in position. These will then govern the smallest torque that causes any motion at all. For example, consider an assay balance consisting of an arm supported on a fulcrum of two hemispheres, Figure 5.9. The spheres support both the weight of the arm mass (plus any pans and indicating instrumentation), m_o, and the weight of the mass to be measured, m. The fractional change in mass that can be detected by the balance i.e the ratio of the minimum change in mass, δm, to the nominal mass, is

$$\frac{\delta m}{m} = \frac{2\mu_r m_o}{Lm} + \frac{4\mu_r}{L} \tag{5.19}$$

The first term on the right hand side causes the most serious limitation of the balance. Note that it depends on ratio of the mass of the balance arm to the mass to be measured. The second term is the limit on sensitivity imposed by the load of the mass itself. There is a factor of 4 in this term because placing a mass in one pan requires that an equal mass be placed on the opposite side of the balance. Taking, for example, an arm of mass 1 kg and length 200 mm, and a typical rolling resistance in the region of 10^{-5} m, the resolution with which a 100 g mass can be measured is 1.2 mg. This will be quite adequate for many purposes, but will become a severe limitation for very high precision systems. Under such conditions, flexure pivots are often used in place of knife edges, Speake, 1987.

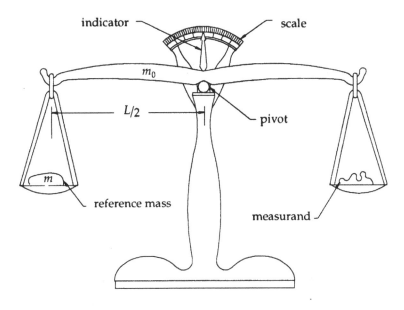

Figure 5.9 An assay balance

5.3.3 The sliding rotary coupling

In many precision applications it is convenient to provide controlled displacements by driving a micrometer or similar leadscrew system. A simple and precise automated drive can be constructed from standard items by coupling an electric motor to a micrometer drum. Often the drum is integral with the anvil and will itself displace as it rotates, requiring the coupling to expand or contract. For applications in which a relatively small contraction is required, the bellows or folded cross couplings shown in Figure 5.7 will provide sufficient 'give' to accommodate the change. A high immunity to thermal changes, extremely low backlash and low mass are the key advantages of such devices. However, as the amount of axial movement increases it becomes necessary to insert a sliding coupling to provide a complete and theoretically unlimited axial freedom.

One form of telescopic coupling effective at light loads incorporates a simple slider as shown in Figure 5.10. It consists of a rod rigidly attached to the motor drive that can freely slide inside a tube that attaches to the driven stage. Relative rotation between the two is prevented by pins that run along a pair of slots machined out of the tube. It is a variant of the spline shaft concept used in power transmission, revised to minimize axial friction while maintaining good rotational stiffness. A further reduction of the frictional force between the tube and rod may be achieved by using rolling bearings at the contact interface.

149

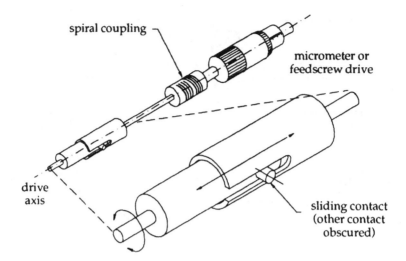

spiral coupling

micrometer or
feedscrew drive

drive
axis

sliding contact
(other contact
obscured)

Figure 5.10 A sliding rotary drive

Angular misalignments are isolated by the use of a spiral coupling (see Section 5.3). Other design inversions also lead to useful forms. As long as the friction between sliding contacts is kept low, the sliding rotary drive provides both smooth transmission and additional vibration isolation along the drive axis. A disadvantage is that any irregularities along the slot will cause a pitch error between the driving and driven axis that is inversely related to the coupling radius. Increasing the radius to reduce the effect of manufacturing error leads to systems of unacceptably high mass. All couplings of this type must cause some backlash since spring loading or elastically averaging several sliding contacts would involve a large increase in the axial friction.

5.4 Case study: Irving Langmuir's surface tension instrument for the determination of the molecular sizes of fatty acids

This instrument has been chosen as a case study not only because of its novel use of a knife edge coupling but also because it demonstrates fundamental characteristics common to most ultraprecision instrumentation. It arose from the work of Irving Langmuir, in the early twentieth century, to determine the elastic properties of thin films. Investigations carried out around that time, mainly by Agnes Pockels and Lord Rayleigh, had suggested that oil films on water might be molecularly thin, as was confirmed by Katharine Blodgett in 1934. Langmuir made a large contribution in this field of study with a monumental series of papers 'on the fundamental properties of solids and liquids'. The majority of his results were obtained using the sensitive balance which is shown schematically in Figure 5.11, Langmuir, 1917. Previous researchers had tried to measure surface tension forces in the presence of films and had been quite successful with a very fine force resolution of the order 0.1 mN being readily achievable, Rayleigh, 1899. The new instrument was developed to provide further increased accuracy and the improved design had

"... the marked advantage is that it affords a highly sensitive and accurate *differential* method of measuring slight changes in the surface tension of water".

The water was held in an enamelled tray measuring approximately 60 cm by 15 cm. Above this a knife edge rests on a glass flat which is attached to a rigid support body, mainly removed from the figure for clarity. The knife edge carries a balance arm which has a counterweight attached to one end and a small pan hanging from the other end. Two glass rods are cemented to the knife edge and extend down into the water through two holes in a strip of paper, previously soaked in a solution of benzine and paraffin, that floats on the liquid surface. The length of the paper is less than the width of the tray so that it can move freely, without touching the sides of tray. Tiny air jets (not shown) could be applied each side of the paper to prevent any films from flowing around the edges of the strip.

Having removed as much of the contaminants as possible from the surface, a 50 mg weight is placed on the pan. This causes the knife edge to rotate moving the paper strip to the left (in the view of Figure 5.11). The strip A is then moved towards B. Any contaminant film still present will lead to the development of a repulsive force between the strips as the separation closes. Thus strip B will be pushed to the right. When it is returned to its original position, the force separating the two strips will be very nearly equal to the weight in the pan scaled by the leverage of the balance. Parallax is elegantly avoided by reading the rule from the reflection in the water surface. The strips are then separated again, the weight removed from the pan and an oil droplet applied to the surface of the water between the strips. This film is compressed by again moving strip A towards B. Weights are added to the pan and the gap closed until B has returned to its equilibrium position. The additional force caused by the presence of the oil is related directly to the difference in weight compared to the first trial. The stress may then be calculated as the ratio of the force to the area of the film. The area presenting to the strips can be calculated by using conservation of volume and measuring accurately the amount of oil added to the surface.

This example has been chosen because it depends critically on near zero resistance at the pivot. It is also directly related to the following chapter on

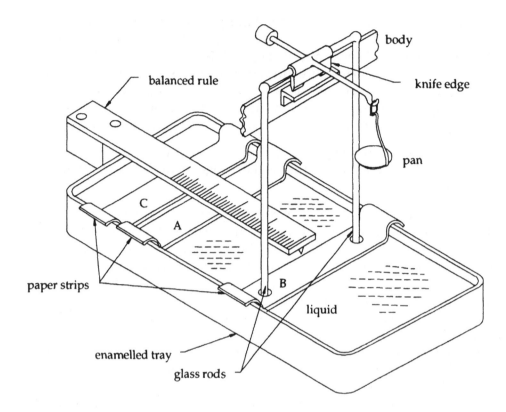

Figure 5.11 Apparatus for the measurement of surface tension of molecularly thin films

leverage, where it will be further discussed. Note particularly that it was possible to calibrate the surface tension against the number of molecules per unit area on the surface using a ruler.

Other interesting aspects of this experiment are the use of nulling to reduce the effect of nonlinearities of the knife edge and differential measurement to compensate for the contaminant film. Fortuitously the nature of the measurement necessarily includes viscous damping which will attenuate vibrations from the surroundings. Interestingly, it was Benjamin Franklin who first reported the ability of oils to spread over the surface of water and unwittingly proposed the spreading of thick layers for the calming of rough seas!

Although not stated specifically in his paper, the resolution of Langmuir's instrument appears to be about 0.015 mN which is an order of magnitude better than that reported by Rayleigh. The main sources of this limitation are: the settling time for movement of the glass rods, the friction between the paper and the glass rods, the accuracy with which the pointer can be aligned with the paper, the precision of the weights and the friction of the knife edge. All of these limitations can be reduced by the introduction of more sensitive measuring techniques except the last one, which is a fundamental limitation of the contact of the knife edge. Neglecting the balance mass, m, Equation 5.19 can be rearranged to give the mass of the balance arm. Inserting the above resolution and using a typical rolling resistance value of 10^{-5} m and an arm length of 50 mm, the mass supported by the knife edge calculates as approximately 0.15 kg. This value seems not unreasonable from views of the equipment shown in his papers.

References

Blodgett K.B., 1934, Monomolecular films of fatty acids on glass, *J. Am. Chem. Soc.*, **56**, 495.

Clark R. A., 1950, On the theory of thin elastic toroidal shells, *J. Math. Phys*, **29**(3)

Eldridge K.R. and Tabor D., 1955, The mechanism of rolling friction. I. The plastic range, *Proc. Roy. Soc., Lond.*, **A229**, 181-198.

Goodman L.E. and Brown C.B., 1962, Energy dissipation in contact friction: Constant normal and cyclic tangential loading, *Trans. ASME; Series E, J. Appl. Mech.*, **29**, 17-22

Halling J., 1959, Effect of deformation of the surface texture on rolling resistance, *Br. J. Appl. Phys.*, **10**, 172-176.

Hamilton G.M. and Goodman L.E, 1966, The stress field created by a circular sliding contact, *Trans. ASME, J. Appl. Mech.*, **33**, 371-375.

Hamilton G.M., 1983, Explicit equations for stresses beneath a sliding spherical contact, *Proc. Instn. Mech. Engrs*, **197C**, 53-59, (there is an erratum related to this paper at the end of this volume).

Heathcote H.L., 1921, The ball bearing: In the making, under test and on service, *Proc. Inst. Automobile Eng's*, **15**, 569

Hocken R. and Justice B., 1976, *Dimensional stability*, NASA Report NAS8-28662, p. 48.

Johnson K.L., 1987, *Contact Mechanics*, Cambridge University Press.

Johnson K.L., 1955, Surface interaction between elastically loaded bodies under tangential forces, *Proc. Roy. Soc., Lond*, **A230**, 531-549.

Langmuir I., 1917, The constitution and fundamental properties of solids and liquids. II. Liquids, *American Chem. Soc. J.*, **39**, 1848-1906.

Lubkin J.L., 1951, Torsion of elastic spheres in contact, *Trans. ASME, Series E, J. Appl. Mech.*, **18**, 183-187.

Mindlin R.D. and Deresiewicz H., 1953, Elastic spheres in contact under varying oblique forces, *Trans. ASME; J. Appl. Mech.*, **75**, 327-344.

Lord Rayleigh (J.W. Strutt), 1899, Investigations in capillarity:– The size of drops.– The liberation of gas from supersaturated solutions.– Colliding jets.– The tension of contaminated water surfaces.– A curious observation, *Phil. Mag.*, **48**, 321-337.

Speake C.C., 1987, Fundamental limits to mass comparison by means of a beam balance, *Proc. Roy. Soc. Lond.*, **A414**, 333-358

Stanley V.W., Franks A. and Lindsey K., 1968, A simple ruling engine for x-ray gratings, *J. Phys. E: Sci. Instrum.*, **1**(2), 643-645.

Tabor D., 1955, The mechanism of rolling friction. I. The elastic range, *Proc. Roy. Soc.Lond*, **A229**, 198-220.

Wahl A.M., 1963, *Mechanical Springs*, McGraw-Hill Book Company Inc., London, chapters 5 & 11.

Young W.C., 1989, *Roarks Formulas for Stress and Strain* (6th ed.), McGraw-Hill, London.

6

LEVER MECHANISMS OF HIGH RESOLUTION

This chapter looks at methods for leverage for the purposes of instrumentation and measurement. The term 'leverage' is taken very widely to encompass all scaling devices that help improve the control of property values. Mostly here this involves means of translating a macro-scale action to a micro-scale event. Since the major part of this book is concerned with actuation and measurement, high precision motion levers are discussed in detail, split for convenience into separate categories of linear and angular techniques. The general concepts of 'gearing' through elastic elements or friction drives are also covered. As is most clear from optical or chemical levers, some mechanisms are useful only for measurement while others can also provide actuation. Both are examined in terms of underlying principle, with no attempt to separate the two according to this classification.

Introduction: Generalized leverage

We start this chapter by examining the capabilities of a few systems that are not always regarded as examples of levers. Looking back to the case study in Chapter 5 there was a need to deposit a molecularly thin oil film of known molecular volume onto the surface of a water trough. A brief calculation indicates the difficulty of this task. Imagine simplified long molecules of diameter 1 nm and length 3 nm (dimensions typical of stearic or cerotic fatty acids) that is to be compressed into a surface area of 150 mm by 100 mm. Approximately 1.5×10^{16} molecules will be required, but the volume occupied by the oil will be only 0.045 mm^3 (or microlitres). Because, for example, of surface tension effects, small fractions of a microlitre are very difficult to dispense with any accuracy. To overcome this problem Devaux, 1913, used a dilute solution (1:1000) of the oils in high purity benzene which will evaporate leaving only the oil. The solution can be made up in large quantities to high accuracy. Assuming that molecular weights have been previously measured and the mixing is uniform, the total number of molecules per unit volume can be readily calculated. This is a rather simplified presentation of a complex process but serves to illustrate the wide variety of techniques that can be considered as leverage. In this case evaporation is being used to effect a *mass lever* and ultimately to enable an evaluation of the size of molecules using a conventional ruler.

The *liquid* lever used to orient instrument components relative to the gravitational field of the Earth, which provides the actuation force, also exploits surface tension to maintain a consistently shaped bubble. The principle of operation of the common spirit level is well known and requires little explanation. A cross-section of a typical spirit level tube is shown in Figure 6.1.

Ideally, the spirit level bubble will rise always to the top of the tube because the gravitational forces on the surrounding, denser, fluid are stronger. In this instance the sensitivity, S, in, say, metres per radian of tilt is simply equal to the radius of curvature, R, of the tube. Its angular resolution depends on the precision to which the relative motion of the bubble can be measured. Using a very large radius of curvature gives high sensitivity to high precision gauges: approximately 100 m is needed to provide 0.5 mm per arc second. Care must be taken to avoid any bending of the tube, whether by thermal disturbance or mechanical strains when mounting the level on an instrument. The effect of distortion can be illustrated by considering a tube of length L, say, for which the total sagittal depth for a large radius of curvature is approximately

$$D = \frac{L^2}{4R} \tag{6.1}$$

For a tube of radius 100 m and length 100 mm the sagitta is about 25 μm. If we want an accuracy of 1% then it must be controlled to better than 250 nm!

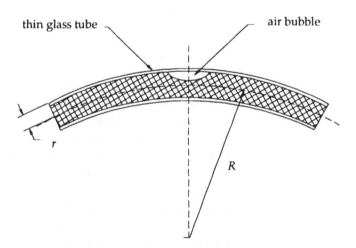

thin glass tube — air bubble

r

R

Figure 6.1 Cross section of a typical spirit level tube

Clearly, dimensional control becomes a major limiting factor associated with this type of measuring device.

Every keen fisherman and vinter will know of a simple method for the amplification of very small forces. Commonly known as the float, it can be considered an Archimedes lever because of its mode of operation. A float will be in equilibrium in a constant gravity field when the volume of displaced liquid equals the combination of its weight and any externally applied vertical forces. Leverage is achieved by ensuring that the part of the float protruding through the liquid surface is very thin and displaces little fluid for a given vertical movement. For a circular prismatic rod, the change in upward force, ΔF_u, due to a vertical displacement of the float, Δz, is found from

$$\Delta F_u = \rho \Delta V = \rho \pi r^2 \Delta z \tag{6.2}$$

where ρ is the *liquid* density and r the radius of the rod. A simple calculation shows the sensitivity and accuracy that can be achieved. A float, submerged in water so that only a rod of diameter 2 mm protrudes from the surface will have a sensitivity of 3.14 mN m^{-1}. Measuring displacements in the micrometre range is not particularly onerous, so nanonewton resolution seems plausible. Accuracy will probably be limited by inconsistency in the rod cross-section. Over moderate distances dimensional accuracy to around a micrometre is realistic. Hence for a displacement of around 1 mm, ΔV should be measured to at worst a few parts in 1000. It is relatively easy to obtain pure water with density known to around one part in 1,000 so calibration errors from this source can be made small. Consequently forces in the order of micro-Newtons should be readily measurable with less than 1% uncertainty. However, in practical situations there are severe problems in approaching this performance due to high compliance in the plane of the liquid, the difficulty in maintaining float stability and vibration sensitivity.

As with many other systems, the float requires the position of a surface to be measured with high sensitivity while minimal forces are imposed on it. A common technique for this measurement is to use an optical lever. Often rather complex in its physical implementation, the principle of operation of the optical lever is very straightforward. A collimated light source is reflected from a mirror and changes in position of the reflected beam due to rotation or translation of the mirror are monitored by a photodiode, or merely by eye against a scale. If the mirror rotates by a small angle φ the reflected beam deflects through 2φ and the lateral shift at a detector placed a distance D from the mirror will be $2D\varphi$. A schematic diagram of a simple optical lever for angular measurement is shown in Figure 6.2a. A beam of light is projected onto a smooth surface that reflects it onto a photo detector array usually consisting of two or four closely spaced detectors. If the beam is incident upon all detector plates with equal intensity, then the outputs will be matched. Fine adjustment of the beam position is provided by an optical micrometer. This consists simply of a flat glass plate of refractive index, n, and thickness, t, placed almost normal to the beam. Light incident at an angle ζ will emerge from the plate parallel to the incident beam but displaced by an amount, δ, which for small angles of incidence is approximated by

$$\delta = \zeta t \left(1 - \frac{1}{n} \right) \tag{6.3}$$

Initially, the optical micrometer is either manually or automatically adjusted to balance the outputs from all the detectors, indicating that the beam is central to the array. Subsequently, if the mirror rotates slightly a suitable rotation is applied to the plate to re-centralize the beam. Given the refractive index and

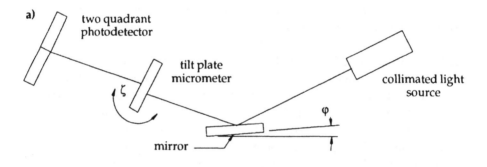

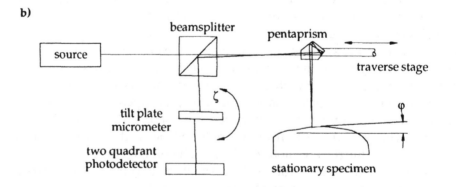

Figure 6.2 Typical optical lever configurations for: a) angular and b) surface profilometry measurement

thickness of the glass plate, the change in its angular orientation, ζ, allows the change in angle of the mirror to be determined. This technique actually involves two lever principles: the primary lever converts a small rotation to a larger displacement and the optical shearing plate converts a small displacement to a larger rotation. It also utilizes a null detection in order to minimize systematic errors.

The layout shown in the figure suffers the disadvantage that it is sensitive to linear displacements of the mirror due to the angle between source beam and photo detectors. This angle can be reduced by placing the two as close together as is practicable and/or moving the mirror far away. In the limiting case, the source and detector may share a common optic axis by means of a beam splitter. By then sharing also the collimating optics, we arrive at the well-known autocollimator instrument. By placing grids in the paths of both the incident and reflected beams, Jones and Richards, 1959, were able to produce optical levers giving a resolution of better than 1 nrad.

Figure 6.2b is a schematic representation of the profilometer of Ennos and Virdee, 1983, that uses optical leverage to measure local surface slope. The profile is reconstructed by integrating the slope at fixed increments along the specimen surface. Of particular note is the pentagonal prism that is traversed above the specimen using a conventional ball bearing slideway. It is a geometric property of a pentaprism that the incident beam is rotated through an angle of $90°$ irrespective of small changes in its orientation. Noise in the slideway used to traverse the prism will not be monitored by the detectors. This effectively takes the slideway out of the measurement loop of the system.

A point to note with all the above examples is that the leverage cannot be readily inverted. For example, it is difficult to imagine a system that moves a bubble to generate a fine angular displacement and the magnitude of displacements that would be feasible by changing the angle of incidence of a beam of electromagnetic radiation are likely to be extremely small. Plausibly, radiation pressure, being proportional to the energy density of the incident and reflected light beam, could be significant when using modern high-powered lasers. Indeed, Jones and Leslie, 1978, obtained twists of a suspended mirror of up to 100 µrads by this method. Such instances are the exception and, in general, mechanical levers are required for actuation.

This book is concerned primarily with mechanical systems and the rest of this chapter will concentrate on the magnification and attenuation of displacements by mechanical means. For convenience we differentiate between angular levers, in which the desired output is a rotation, and linear ones, which cause a displacement. At a first glance it may seem obvious to use a mechanical lever for all types of micro-manipulation. However, there can be a number of disadvantages, typically caused by scaling from one dimension to another, and these are discussed in the following section.

6.1   The effects of levers

There is obvious appeal to the precision engineer in producing components of 'normal' engineering proportion by standard machining techniques and then simply applying levers to suit them to ultra-precision applications. This follows the design philosophy of attaining high precision with minimum resort to high precision, and therefore high cost, manufacturing processes. Unfortunately, there are a number of pitfalls associated with this approach. A simple example is the use of a standard planetary gearbox for the reduction of angular rotations from a stepper motor. Although gearboxes having reduction ratios of many thousands are available, they are not generally used for fine angular positioning because backlash in the gear teeth, expressed as the angle of rotation for each gear, is magnified in proportion to the reduction ratio. Methods for the reduction of these effects will be discussed in Section 6.2. The magnification of small scale nonlinearities is a problem common to many lever systems. Typical causes of nonlinearity are friction in bearings and sliding surfaces, thermal expansion and thermoelastic effects, incomplete elastic recovery on unloading, ineffective lubrication, surface finish on datums and manufacturing errors. Some are often only manifest if the lever is used for cyclic, reciprocating applications, where they appear as hysteresis, and so may be less serious in applications requiring continuous, unidirectional motions.

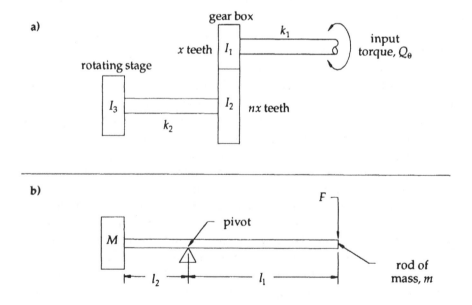

Figure 6.3   Simple lever mechanisms; a) Gearbox; b) Lever arm.

Whether using leverage to amplify or attenuate motion, there is almost always a dynamic penalty to be paid. This can be seen intuitively. Imagine trying to use a door to drive a slight angular motion of a small instrument stage by applying a relatively large linear motion to its handle. Acceleration will be resisted by the combined inertias of both the stage and the door, with the latter totally dominant. Conversely, using a small angular rotation by pushing near to the hinges of the door to provide a larger displacement is an obvious example where dynamic response will be reduced. If not convinced, try to slam a door by pushing near to the hinge. For a more rigorous analysis consider the simple gear and lever mechanisms illustrated by Figure 6.3.

Figure 6.3a represents a simple reduction gear consisting of an input shaft of stiffness k_1 attached to a disc gear having x teeth. This meshes with a further gear of similar width and nx teeth. If the gears are very rigid and the effects of manufacturing errors can be ignored, the ratio of angular position between input and output shafts is exactly n. The two gears have inertias (second moments of mass, or moments of inertia, since the motion is rotary) I_1 and I_2 respectively. Connected to the output gear is another shaft of stiffness k_2 which drives a rotary stage of inertia I_3. The kinetic and potential energies of this system can be easily derived as

$$T = \frac{1}{2} I_1 \dot{\theta}_1^2 + \frac{1}{2} I_2 \dot{\theta}_2^2 + \frac{1}{2} I_3 \dot{\theta}_3^2$$

$$= \frac{1}{2} I_1 \dot{\theta}_1^2 + \frac{1}{2} I_2 (n\dot{\theta}_1)^2 + \frac{1}{2} I_3 (n\dot{\theta}_4)^2$$

$$U = \frac{1}{2} k_1 (\theta_1 - \theta_0)^2 + \frac{1}{2} k_2 (\theta_3 - \theta_2)^2$$

$$= \frac{1}{2} k_1 (\theta_1 - \theta_0)^2 + \frac{1}{2} k_2 n^2 (\theta_4 - \theta_1)^2$$

(6.4)

where $n\theta_4 = \theta_3$.

Substituting Equations 6.4 into Lagrange's Equation yields

$$(I_1 + n^2 I_2) \ddot{\theta}_1 + k_1 (\theta_1 - \theta_0) + k_2 n^2 (\theta_4 - \theta_1) = Q_\theta$$

$$n^2 I_3 \ddot{\theta}_4 + n^2 k_2 (\theta_4 - \theta_1) = 0$$

(6.5)

The first term of Equation 6.5 indicates that the apparent inertia of a system element is always increased from its basic value by the square of the gear ratio through which it is driven. If the teeth can be considered small in relation to the radius of the gear its inertia can be approximated by that of a circular disc and proportional to the square of its radius. To mesh properly the teeth must be the same on each gear, so the radius of gear 2 is n times that of gear 1. Thus, the inertia of the gear train seen by the drive will increase in proportion to the fourth power of the gear ratio. Analysis of the simple lever of Figure 6.3b leads to the same conclusion. The introduction of any leverage or gearing may severely restrict the dynamic response of a system. There will be an associated reduction in its resonant frequencies, which may be of even greater concern.

Another problem commonly found with high ratio precision levers is that of 'lost motion'. This can be simply illustrated by the lever arm of Figure 6.3b. A small input displacement at the right hand end, requiring a force F_1, will be attenuated by a factor l_2/l_1 (= n, say). The drive force F_1 will similarly be magnified so that the force at the left hand end, F_2, is given by nF_1. These forces must be supported by a reaction force at the pivot of magnitude $F_3 = F_1(1 + n)$. All real pivots have a finite stiffness, λ, so the fulcrum displaces in the direction opposite to that of the output motion. As a result the final position of the left hand end will have moved from its original position by a smaller distance than expected. The error in the displacement, or lost motion, is

$$\varepsilon = \frac{F_1(1 + n)}{\lambda} \tag{6.6}$$

This term may become significant if either high forces or large lever ratios are used. It poses a particular danger if relatively stiff systems are to be moved with high precision.

Notwithstanding the disadvantages indicated above, levers often present the cheapest solution for precise manipulation using standard engineering components. A range of designs and applications are considered in the following sections concerned respectively with the categories of angular and linear devices.

6.2 Angular levers

We discuss angular levers before linear ones because angular adjustment to high precision is often quite simple and relatively insensitive to environmental disturbance. This is because an instrument constructed completely from a single homogeneous, isotropic material will be naturally insensitive to slow thermal changes to the extent that, although the whole instrument or machine may expand, relative angles between components will remain the same. Thermal gradients introduce only a second order effect. This can be simply illustrated by considering a long lever of length L pivoted at one end and displaced a small distance, s, at the other by a micrometer. The angular change is simply s/L. If the lever, micrometer *and* base are constructed from the same material, a change in temperature T changes the distance of the micrometer from the pivot to $L(1 + \alpha T)$ and its spindle extension to become $s(1 + \alpha T)$. The ratio and so the angle of the lever is unchanged. It should be stressed that this happens because each part of the complete measurement loop, including the supporting base, changes length in scale. As the simple lever has already been considered in the previous section, three techniques for angular adjustment will be examined; the pulley, the wedge, and gears.

6.2.1 The pulley drive

A rotation can be relatively precisely transformed to a linear motion by use of the ribbon drive, see Figure 5.6. Conversely, if the rotation axis can be fixed, the ribbon may be moved to effect a rotation. In this instance the angular rotation per unit displacement is simply the reciprocal of the radius of the spool. More interestingly, for applications in which angular rotations are small and the axis of rotation need not remain stationary, the ribbon can be wound around a spool comprising two different radii, see Figure 6.4. To rotate the spool, one end of the ribbon is fixed and the other undergoes a small displacement, x. The bobbin

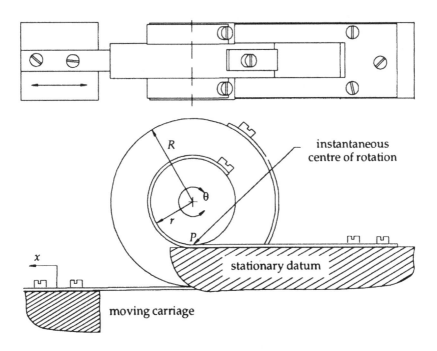

Figure 6.4   A linear to angular motion magnifying lever

is constrained by the pair of fixed ribbons to rotate about the instantaneous centre of rotation, P, located at the point of initial contact of those ribbons on the spool. The total drive motion is made up of some unwinding of its ribbon from the spool and some translation of the centre by the unwinding of the fixed ribbons. The sensitivity of angle of rotation, θ, due to the drive displacement, x, is given by

$$\frac{\theta}{x} = \frac{1}{(R - r)} \tag{6.7}$$

The denominator can be varied over a large range of values to provide high magnifications, with no sliding components to induce wear or hysteresis. Attaching a pointer to bobbin yields a sensitive dial indicator comprising few components that is relatively inexpensive and simple to construct, Elliot and Home-Dickson, 1959. The same principle can be used to provide fine motions by means of a long torsion arm. Another variant is discussed below.

6.2.2 The wedge

A wedge pushing a suitably constrained kinematic contact provides a simple method of scaling down a displacement. It is particularly effective when used with a lever to achieve fine angular adjustment from a relatively coarse linear positioning device. A typical system is shown in Figure 6.5. An adjustable platform rests on a pivot at one end and, using a suitably shaped contact, a wedge at the other. The wedge can be driven along a slideway approximately parallel to the platform, or lever. This particular mechanism was used for levelling specimens in the stylus profilometer of Lindsey *et al.*, 1988. It has a pivot consisting of two 10 mm radius hemispheres and a wedge angle of approximately 1°. The platform is maintained in position by applying a downward force to it by leaf springs (not shown). Slip between the wedge and platform is ensured by locating these springs nearer to the hemispherical supports than to the contact point of the drive. The micrometer is manually adjusted, so all the loop components are constructed from the glass ceramic 'Zerodur' to reduce effects of thermal gradients. Using a simple rotating micrometer of 0.5 mm pitch it has proved possible, with visual feedback, to achieve a resolution in angular adjustment of better than 50 nrads (0.01 arc seconds). As with the ribbon device, adjustment of the stage changes slightly the position of the axis of rotation. This may be undesirable in many applications. It is worth noting in passing, that a simple lever and wedge drive, again constructed from Zerodur, was produced to adjust a small probe towards a surface. Using a 0.5 mm pitch thread it was possible to increment towards the surface in steps of approximately 10 nm directly from the micrometer.

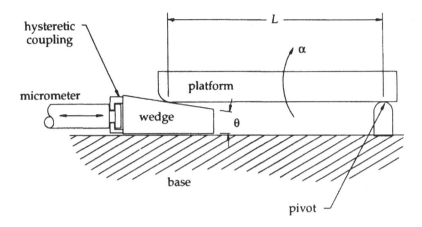

Figure 6.5   A simple wedge type lever mechanism

Another potential disadvantage of wedge systems is that their performance is strongly influenced by manufacturing tolerance on precision as the magnification is increased. To achieve greater resolution, either the length of the platform must be increased or the angle of the wedge reduced. The former is likely to introduce thermal instability as the contact point will climb up the wedge upon thermal expansion of the platform. The latter relies on the angular accuracy between the wedge and the axis of motion of the slideway.

6.2.3   Angular gearboxes and friction drives

All the angular lever mechanisms discussed so far suffer the drawback of being capable of only a comparatively small angular displacement. For positive action, continuous rotary motion some form of sliding system is required. Most commonly, toothed gears are used, as was assumed in the discussion at Section 6.1. However, they introduce noise into a system due to the intermittent meshing and releasing of individual gear teeth, which makes them unattractive for high precision applications. Special methods are needed to avoid backlash in reversing applications, without introducing excessive internal stresses. Tooth to tooth stresses are always high and there is inevitably a reciprocating sliding motion between tooth faces that makes them vulnerable to scuffing types of wear. These also are undesirable. For large and variable loads where a positive drive is essential this type of gearing may be the only viable method and noise from the gearbox must then be removed using vibration isolation techniques. In applications where the loads are smaller and likely to remain relatively constant, friction drives may be used. They are often the optimum solution for rotary gearing having a moderate reduction ratio of between unity and 50:1. For reduction ratios of greater than this, more involved techniques must be used. A bonus with most friction drives is that a continuous range of gear ratio can normally be obtained. Since there are many texts that cover the design of conventional gearing, and this is certainly not the place to study gear tooth design in detail, we shall concentrate on the friction driven, smooth motion devices.

A simple reduction gear can be constructed by terminating each of two parallel shafts by a conical solid of revolution, with their apices pointing in opposite directions. The cones are coupled by an idler wheel, or flexible belt, which may freely rotate while pressed against them, Figure 6.6a. Under conditions of low torque transmission the slip between faces will usually be less than 1%, so the gear ratio may reasonably be assumed to be the ratio of the radius of the driver to that of the driven cone at the point of contact, r_1/r_2. Moving the idler along the faces of the cones provides a continuous variation of this ratio, in practice from around 1:3 to 3:1. Alternatively, a flat disc can be connected to the end of the driving shaft and a roller be rigidly attached to the driven shaft, the rotation axes being perpendicular, see Figure 6.6b. Again, the

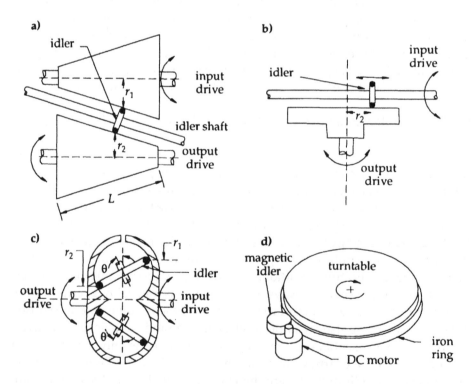

Figure 6.6 Rotary speed converters; a) Cone and roller, b) Disk and roller, c) Toroid and roller, d) Free roller and disk.

roller can be positioned along the disk to vary the speed conversion ratio from about unity up to 100:1 or more. Two more of the many variations on this theme are shown in Figure 6.6c & d. The first has idler disks squeezed between two toroids. The gear ratio is adjusted by changing the angle of the axes of the idler disks, θ. The available range is typically from 1:5 to 5:1. This design has the advantage that the torque about the drive axis is balanced by the symmetrical arrangement of idlers, permitting the use of large contact forces. For this, and other reasons, such gearboxes are to be found in large pay-load vehicles and other applications where it is desired to transmit large forces over a continuously variable gearing range. The final design uses a free running idler that is also a permanent magnet. Adhesion is maintained by using ferromagnetic materials for the driver and driven rollers. Its advantage is that there is no bearing noise introduced by the intermediate element that is there only to ease the matching across the drive components. Using such a system, Teragaki and Kobayashi, 1987, were able to drive a 300 mm record player turntable from a 3,000 rpm DC motor having a spindle diameter of 3 mm (reduction ratio of 100:1) without introducing any audible noise.

Differential techniques are capable of generating high reduction ratio friction drives using relatively simple components. A particularly interesting example is presented in a patent by Koster *et al.*, 1986. As illustrated by Figure 6.7, an intermediate disk is attached to a rigid and stationary base via a flexible diaphragm. One face of the disk is rigidly attached to the drive shaft by a crank that forces it to trace out a conical path of half angle α. The opposite face is of.

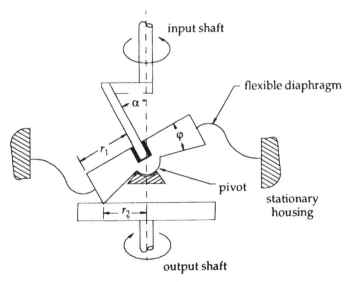

Figure 6.7 Sectional view of the friction disk speed converter

conical section of half angle β and is freely pivoted about the centre line. The outer rim of the conical surface contacts against a flat disk that is attached to the output shaft of the gearbox. The input and output shafts are colinear. Accurate analysis of this drive must account for slip in the contact zone and movement of the instantaneous axes of rotation. Simplifying by ignoring such effects, consider only the length of the locus of contact for each disk. Clearly, for the driven disk, one rotation of the drive shaft will cause the full circumference of the outer radius of the cone, r_1, to traverse over the surface of the driven disk. However, the radius of the contact circle on the driven disk, r_2, is smaller by an amount

$$r_2 = r_1 [\cos(\alpha) - \sin(\alpha)\tan(\varphi)] \qquad (6.8)$$

Consequently, the additional rotation that the driven disk must undergo during each input rotation to provide compatibility provides a gearing

$$\frac{\omega_{out}}{\omega_{in}} \approx \cos(\alpha) - 1 - \sin(\alpha)\tan(\varphi) \qquad (6.9)$$

A more precise value for the ratio, $-\tan(\alpha)\tan(\alpha + \varphi)$, can be derived by considering the instantaneous centres of rotation about the contact region. However, for the values for α and φ of 0.5° and 13° quoted by Koster *et al.*, 1986, both models give a reduction ratio of –500:1 to within 2%. The negative sign indicates that direction of the output shaft opposes that of the input. Another distinct advantage of this type of gear is that the diaphragm allows the rotary drive to pass into a vacuum system without the need for any sliding seals.

A related approach uses gear teeth machined onto each side of a flexible cylindrical membrane. By having a different number of teeth on one side to that on the other the ratio of the input to output can be arranged to be the fractional difference. The variations on this theme are commonly called harmonic drives. Although capable of high gear ratios and of transmitting large forces, the majority of such gearboxes examined by the authors have exhibited noise levels that were unacceptable for low vibration, high precision applications. There are other friction driven variations on the harmonic drive, some of which may have a central diaphragm for vacuum applications.

6.2.4 Balanced torque actuation

There is a not infrequent need with elastic mechanisms to provide small torques, typically for error compensation in instruments, with minimal variation in lateral force. Although Chapter 7 is dedicated to actuators, this particular requirement is best treated here.

Redistributing the mass within a closed system may alter the gravitationally induced torque needed at its supports, while leaving the total weight unchanged. In principle a pure torque drive could be so achieved. The first monolithic x-ray interferometer, Hart, 1968, used this technique. A schematic illustration of its implementation is shown in Figure 6.8.

A flexible chain hangs from attachments at each end of a tilt platform and wraps around a pulley with an overhang at each side. Rotation of this pulley increases the length of chain overhanging to one side, and reduces it equally on the other, resulting in a torque that is proportional to difference in overhanging lengths. The mass supported from each end of the platform is proportional to the length of chain from the connection at the end of the platform to the point of maximum sagitta. The advantage of this type of adjuster is that the total mass of chain supported by the platform remains constant upon adjustment. The resulting relatively pure torque will not be effected by uniform thermal expansion of the chain. Expansion of the pulley does influence the torque, but the effect is likely to be negligible under normal laboratory conditions. An obvious drawback of the mechanism is its low dynamic range and susceptibility to vibration caused by the low natural frequency of the pendulous chains.

Although gravitational forces are extremely constant, they are difficult to utilize for dynamic applications. Balanced pairs of other force actuators operating against levers may therefore be needed. Chapter 7 examines various approaches to actuation, so here one example will suffice. Conventional simple

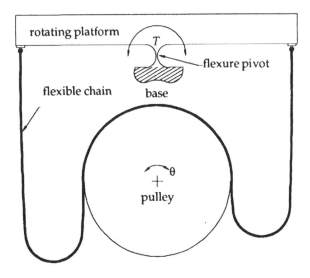

Figure 6.8   Hanging chain torque adjuster

flexure mechanisms for linear motion, see Chapter 4, have relatively low torsional stiffness in the pitch axis. It may prove impracticable to align the drive sufficiently well to avoid significant parasitic torques in this axis, so a compensation system would be sought. (We note that this problem is closely allied to that of deliberately constructing a linear plus pitch two-freedom device by exploiting the low torsional stiffness.) One possible solution is to drive the mechanism as shown in Figure 6.9.

Rather than mounting a single central drive onto a monolithic parallel spring, two permanent magnets with their axes parallel to the drive axis, z, are placed on a cross-beam symmetrically to the spring centre-line, a distance L apart. Surrounding each of the magnets, but without physical contact is a solenoid coil. For suitably positioned magnets, a force will be exerted on them proportional to the current passing through their coil, Smith and Chetwynd, 1990. The sum of these forces causes a linear displacement of the spring mechanism. Because of the offset of the magnets from the central axis of the spring, the difference between force-offset products yields a torsional actuation. In principle the couple generated by one solenoid/magnet actuator can be compensated by applying an equal couple from the other to produce a pure rectilinear motion. Alternatively, the linear force can be compensated to provide a pure couple with, obviously, a continuous variation between these extremes being possible to give a true two degree of freedom actuator.

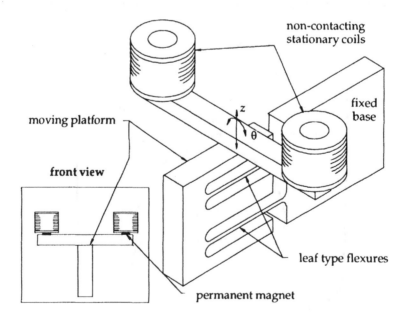

Figure 6.9   A two degree of freedom electromagnetic actuator

Although simple in principle, there are a number of practical complications with this device. The lever arms will not be identical, so even with well-matched coils the current drives will need different calibrations to match torques. It is extremely difficult to wind identical coils and the inconsistencies often encountered between the intrinsic magnetization of commercially available permanent magnets may result in the force/current characteristics of the two actuators being significantly different. These effects could be compensated by the use of tracking current drives with inbuilt corrective modulation. Other parameters such as the thermal coefficients of the permanent magnets (approximately 0.04% K^{-1} for samarium cobalt and 0.08% K^{-1} for neodymium iron magnets) may introduce nonlinear behaviour that is less readily dealt with in applications of the highest accuracy.

6.3 Linear levers

The simple see-saw (or teeter-totter) type of lever, already discussed in Section 6.1, may be used for the transmission of linear motion over small displacements. More uniform motion, or somewhat greater range is obtained by driving it through wobble pins. The reduced dynamic response of this approach has already been observed. However, other features of levers used for linear displacement pose more severe limitations on their precision than is the case with angular motion. Thermal expansion will be added vectorially into the output displacement. Also, upon thermal expansion, the line of action of the lever will move causing either transverse forces on wobble pins or sliding at the drive interface. As previously mentioned, if all loop components are constructed from similar materials, slow thermal drifts will be compensated. Monolithic designs are attractive in this regard. Pivot accuracy, see Chapter 5, and lost motion must also be considered. In these respects, also, flexure pivots are attractive because of their consistency. Thus we concentrate on examples of monolithic, spring pivot designs.

Figure 6.10 shows some typical monolithic lever geometries. Figure 6.10a is a simple lever constructed using notch flexures as pivot points. The ideal relationship between the linear motion at the input and that at the output is simple proportion in the ratio a/b. Unfortunately, as with many other lever mechanisms, it exhibits the most general form of rigid body motion, a combined rotation and a translation. A pivot is provided at each end of the lever to decouple the rotation from the translation. With flexure hinges there must be a torque associated with this motion which will ultimately result in parasitic twists. These can only be reduced by introducing wobble pins or other rolling contact couplings, examples of which are presented in Chapter 5. Figure 6.10b and c show alternative geometries where the input and output motions are set

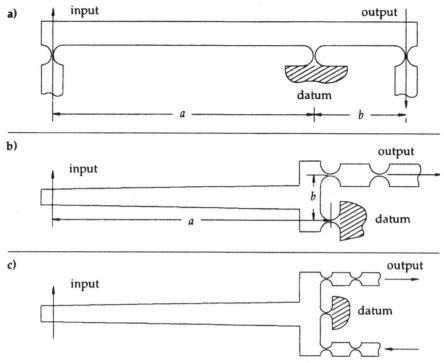

Figure 6.10 Typical lever mechanisms

at right angles. They incorporate pairs of notch hinges in the drive to provide further decoupling of parasitic forces. Figure 6.10c is an example of a push-pull mechanism. If the distances from each drive arm to the datum pivot are equal then their relative displacements will be of identical magnitude in opposite directions. Jones, 1952, used a similar mechanism constructed from leaf springs for the adjustment of an optical slit with undetectable hysteresis and a resolution of better than 50 nm. The datum pivot was also on a leaf spring and could be moved relative to the other two flexures. Careful adjustment of the pivot point then enabled the adjustment of the two ratios to compensate for errors of manufacture and construction. High ratio levers are best avoided since their errors tend to rise with increasing magnification, with, perhaps, 10:1 being a realistic limit. Higher ratios are possible by cascading simple levers. Care must be taken when designing more complex mechanisms to ensure that they are not overconstrained. An example of a monolithic linear spring with three simple levers cascaded into the drive is shown in Figure 6.11 (see also, for example, Deslattes 1969, Nakayama et al., 1982). The outer section is a simple linear spring. Within it, the drive displacement is attenuated by three cascaded levers, each pivoted to the base and reducing the displacement by a factor of three to yield a ratio of approximately 27:1. It is likely that lost motion will have significant effect on this ratio. Bars with pairs of notches act as wobble pins between each section. As a design check, Grubler's formula, Equation 3.5, can be applied. The mechanism has ten links and thirteen joints, consistent with a single degree of freedom mechanism which is ideally suited to a single degree of freedom drive, provided that the axes of drive and actuator are coincident.

An alternative, and often more compact, technique for obtaining fine displacements is to cascade two springs of different stiffness. The principle is very simple. A platform requiring fine adjustment is attached to a strong spring to which is colinearly attached another weaker spring. The other end of the weak spring connects to the drive motor or micrometer. As the drive deflects the weak spring, the strong spring is subjected to the same force and so deflects a corresponding, but smaller amount. This concept can easily be extended to angular levers as discussed more fully in the case study of a micro-tensile testing instrument at the end of this chapter. It is a simple matter to design springs of similar overall dimension having up to four orders of magnitude difference in stiffness value. Consequently, is possible to reduce a standard micrometer from a 25 mm full range to 2.5 µm with a corresponding typical resolution scaled from 10 µm to 1 nm.

Typical implementations with elastic linear mechanisms, convenient because they have inherent stiffness, are shown in Figure 6.12. The first consists of two linear spring mechanisms each attached to the fine displacement platform the other ends attached to the base and coupled to a micrometer drive

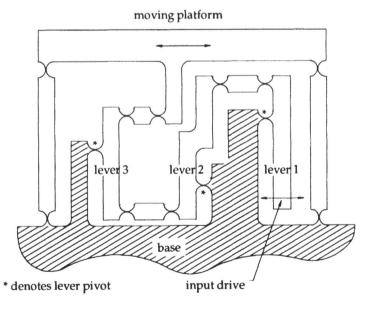

Figure 6.11 Cascade monolithic lever drive for a simple linear spring

respectively. It is equivalent to a mechanism in which two series springs, of stiffness λ_1 and λ_2 respectively, are connected to the base at one end and a micrometer at the other. The ratio of input displacement from the micrometer, x_i, to the displacement of the platform relative to the base, x_0, is given by

$$\frac{x_0}{x_i} = \frac{\lambda_2}{\lambda_1 + \lambda_2} \tag{6.10}$$

For the linear springs of Figure 6.12a it is reasonable to have up to an order of magnitude difference in thickness of the flexures. Because the stiffness is proportional to the cube of the thickness the effective lever ratio ranges from 0.999 to 10^{-3}. Higher ratios can be obtained by attaching thin foil springs, as shown in Figure 6.12b, or any other low stiffness elastic coupling. Achieving a long range with high resolution will require large distortions of the drive flexure which may result in nonlinear behaviour. An advantage of spring to spring leverage is that thermal expansion is scaled by the demagnification and can therefore be ignored in many designs. However, significant errors may

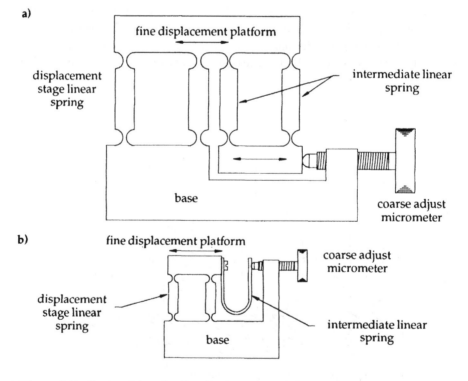

Figure 6.12 Spring drives for fine displacement; a) Compound monolithic linear springs, b) leaf spring drive.

arise from residual stress and 'tin canning' in the drive spring. Thin foils and plates are often produced by cold rolling which involves considerable strains. At constant temperature these are maintained in equilibrium, but as they are subject to temperature variation or large distortion the distribution of stress can alter, resulting in instability. Tin canning is commonly found with thin foil spring mechanisms and is notoriously difficult to eliminate. The examples shown hardly do justice to the large number of mechanisms that operate on similar principles. Combinations of linear and torsional stiffnesses provide many mechanisms involving, for example, pushing against a stiff spring with galvanometer needles, floats, thin wires, or arms connected to twisted tubes. The list is endless and it is left to the reader's imagination to dream up appropriate ideas.

Elastic linear systems, although capable of extremely high resolution, are limited in their range of movement. For longer range traversals, sliding or rolling mechanisms are invariably necessary. Feedscrews are often used for long range fine translation and for applications where adjustments to a few tens of nanometres are required. They are discussed in Chapter 7, but since they involve a principle of leverage a few comments are included here. For precision translations the ordinary feedscrew and nut contains excessive backlash and nonlinear stiffness because it involves sliding contact. To avoid seizure relatively large clearances and light loads must be used. More elaborate screwthreads and nuts can be produced, in particular ball and planetary roller screws are used. These try to overcome the problems associated with sliding contact by introducing rolling motion at the thread faces by interposing ball bearings or a series of rollers between the screw and nut. To increase the stiffness of coupling the contacts should be preloaded. Although backlash is reduced using this method, it is not removed and typically a hysteresis of around 100 nm remains with open loop drives. This may be reduced to a few tens of nanometres under closed loop control, Weck and Bispink, 1991. Further increases in precision may be possible by using a two nut system and incorporating an axial preload. Reducing the number of rolling contacts to two results in the friction drive which is simply a prismatic bar clamped between two rollers. Because this constitutes a complete drive, discussion of it is placed in the actuators section of Chapter, 7.

An interesting feedscrew and nut style of friction drive is shown in Figure 6.13. Rollers consisting of low angle truncated cones are held against a uniform shaft. It can be thought of as a planetary roller drive with infinitesimally fine threads. To maintain the conical rollers in contact with the surface of the drive shaft, their rotary axes must be at inclined at the half angle of the cones relative to the shaft axis. Analysis of instantaneous centres of rotation reveals that the pitch of the drive can be approximated as the product of the length of the cones and their half angle. An easily produced cone might have a half angle of $0.5°$

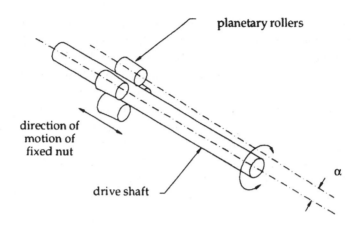

Figure 6.13 Friction planetary roller feedscrew

and length 10 mm, in which case the 'pitch' comes out to be 100 μm. This drive has the advantages that surface finish effects are averaged over the length of contact and that its rolling motion provides superior characteristics to systems using sliding. Similar designs, typically of the dimensions just mentioned, are now commercially available.

6.4   Case studies

The two case studies discussed below are both excellent examples of precision instrument design. They could be used as illustrations in several chapters of this book, but they have particular relevance to many of the ideas presented in this chapter.

6.4.1   The Marsh micro-tensile tester

Marsh's micro-tensile tester was originally developed to investigate the mechanical properties of whiskers, Marsh, 1961. Figure 6.14 gives a schematic representation of the essential features of the instrument. A tensile force is applied to the specimen by means of a lever driven by a torsion wire. A micrometer drive applies twist to the wire and the resulting torque causes the flexure pivot to rotate, imparting a force on the specimen. At this point it is apparent that some force will be 'lost' to the flexure pivot as the lever moves. This error is reduced by moving the specimen holder back to its original position using the nulling micrometer and another lever system to reposition the specimen clamp. Normally a simple two stage lever having a reduction ratio of 100:1 was used for this purpose. For smaller or stiffer specimens, where

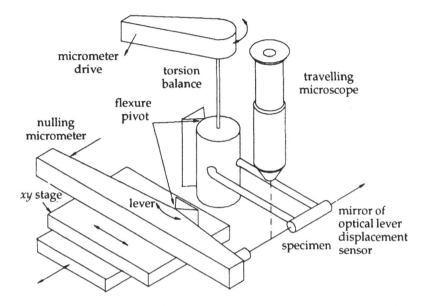

Figure 6.14 Schematic diagram of essential features of micro-tensile testing instrument

a resolution of better than 1 nm may be necessary, the option of a further lever stage provided a ratio of up to 1000:1. The smallest graduation on the vernier micrometer then corresponded to a specimen movement of 0.1 nm. A typical measurement cycle would involve the following sequence of operations.

1. Mount the specimen in the holder by gluing or clamping.

2. Use the x-y stage to adjust for specimen length and to align the jaws of the specimen holder and the clamp the stage. This is done with the aid of the travelling microscope, which can also be used to measure the length of the specimen.

3. After zeroing the optical displacement detector, apply a load through the torsion balance by adjusting the micrometer drive to twist the wire.

4. Adjust the fine adjustment levers using the nulling micrometer until the detector is again reading zero.

5. Now, because the optical gauge has been nulled, the tensile force on the specimen can be computed as a product of the twist on the wire multiplied by the apparent stiffness at the specimen holder (dependent on its torque-arm length).

6. The stretch of the fibre can be deduced from the displacement of the nulling micrometer.

7. Apply a further adjustment to the drive micrometer, as required, and go back to 4.

This design not only incorporates displacement and force levers, both supported by flexure pivots for high repeatability, but also uses an optical lever to provide very high sensitivity to the specimen position detection. A number of other features inherent to it are typical of many good precision instruments. Firstly, the use of nulling with the optical lever means that forces are not lost in the flexure pivot supports since the deflection is the same at the time of every reading. Because the end of the specimen holder to which the force is applied also returns to the same position for every reading, non-linearites in the gauge are unimportant. Simple, but very sensitive techniques can be used without the need for complicated calibrations. Perhaps more importantly, elements within the force loop are all constructed from components having stiffnesses much higher than that of the average fibre and nulling is used to divert the loop away from those components that would otherwise be affected by changes in force. With various lever configurations, this instrument could apply loads from 10 μN to 4 N and detect specimen extensions from a fraction of a nanometre up to 15 mm under constant temperature conditions. Another excellent and more recent example would be the instrument for measuring molecular forces as a function of separation presented by Israelachvili and McGuiggan, 1991.

6.4.2 The Ångstrom ruler

The principles of the x-ray interferometer have already been described in Section 4.6.1, where construction from a monolithic single crystal was justified. Here we are concerned specifically with the design solutions adopted for a particular application by Hart, 1965. One difficulty common to all x-ray interferometers is the reduction of residual twist between the blades to well below 10^{-7} radians. As illustrated by Figure 6.15, a hanging chain torsion adjustment was employed. Under operational conditions it gave an angular resolution of better than 0.001 arcseconds. Such high resolution demonstrates the extreme scaling that can be achieved by pushing or twisting a stiff spring by a small force. The drive mechanism provides another example. The elastic mechanism of the monolith was driven by an in-line micrometer through a U-shaped foil leaf spring. There was a difference of 10^5 between the stiffness of this spring and the mechanism, giving a displacement of 0.1 nm per 10 μm traverse of the micrometer.

The unusual folded flexure of the monolith stage is of interest. This geometry was chosen mainly because it can be manufactured using only a simple diamond slicing machine, which was one of the few facilities available for working in hard, brittle materials at that time (modern machines using high

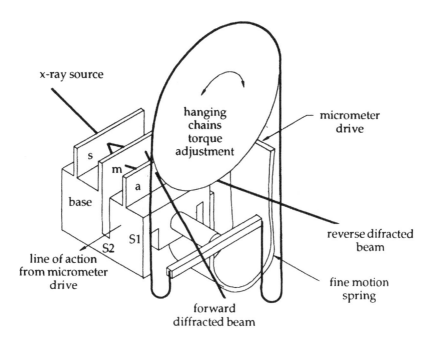

Figure 6.15 Hart's original x-ray interferometer

speed spindles and miniature diamond cutters are capable of producing more intricate shapes in such materials, see, for example, Smith, 1990). The splitter, s, and mirror, m, blades of the interferometer are intrinsic to the base of the monolith. The whole device is mounted onto a rigid support on the underside of this central block. This has the advantage that stresses from the drive through the support are kept far away from the interferometer blades, which remain virtually unstrained. The spring action is provided by weakening the monolith at the positions S1 and S2 by making three simple saw cuts in the block, one parallel to and between the blades, the others at right angles to the first and cut from the underside. The analyser blade is left isolated on a block connected to the base only by the flexures.

Although the flexure spring is of rather unusual configuration and bends in two planes, provided that the drive was positioned correctly, x-ray fringes were observable over displacements of up to 0.1 μm if contrast was occasionally enhanced using the hanging chain adjuster. Backlash in the drive system appeared to be at most 2 pm. Because of the high stiffness of the monolith spring, around 1 mN nm^{-1} and the low mass of the moving components, resonant frequencies were sufficiently high that only relatively simple vibration isolation was necessary in order to achieve this performance.

References

Deslattes R.D., 1969, Optical and x-ray interferometry of a silicon lattice spacing, *Appl. Phys. Letts.*, **15**, 386-388.

Devaux H., 1913, A review of his work can be found in the annual report of the *Smithsonian Institute*, 261

Elliot A. and Home-Dickson J., 1959, *Laboratory Instruments: Their Design and Applications*, Chapman and Hall, London.

Ennos A.E. and Virdee M.S., 1983, Precision measurement of surface form by laser autocollimation, *Proc. SPIE*, **398**, 252-257.

Hart M., 1968, An Ångstrom ruler, *Brit. J. Appl. Phys. (J. Phys. D)*, **1**, 1405-1408.

Israelachvili J.N. and McGuiggan P.M. 1991 Adhesion and short-range forces between surfaces. 1. New apparatus for surface force measurements, *J. Mat. Res.*, **5**(10), 2223-2231

Jones R.V., 1952, An optical slit mechanism, *J. Sci. Instrum.*, **29**, 345-350.

Jones R.V. and Richards J.C.S., 1959, The recording optical lever, *J. Sci. Instrum.*, **36**, 90-94.

Jones R.V. and Leslie B., 1978, The measurement of optical radiation pressure in dispersive media, *Proc. Roy. Soc. Lond.*, **A360**, 347-363.

Koster M.P., Soemers H.M. and Dona M.J.J., 1986, Friction-disk transmission comprising a tiltable disk, *European Patent Specification*, Publication No. 0,224,949.

Lindsey K., Smith S. T. and Robbie C. J., 1988, Sub-nanometre surface texture and profile measurement with Nanosurf 2, *Annals of the CIRP*, **37**(1), 519-522.

Marsh D.M., 1961, Micro-tensile testing machine, *J. Sci. Instrum.*, **38**, 229-234.

Nakayama K., Tanaka M. and Morimura M., 1982, Observation of x-ray interference signals from a two crystal x-ray interferometer, *Bull. of NRLM*, **31**, 3-8.

Smith S.T. and Chetwynd D.G., 1990, An optimized magnet-coil force actuator and its application to linear spring mechanisms, *Proc. Instn. Mech. Engrs.*, **204**, C4, 243-253.

Smith S.T., 1990, The machining of silicon with high speed miniature diamond cutters, *J. Phys. D.*, **23**, 607-616.

Teragaki T. and Kobayashi H., 1987, A new rotary speed converter, *Trans. IEEE*, **23**, 2200-2203.

Weck M. and Bispink T., 1991, Examination of high precision slow motion feed drive systems for the sub-micrometre range, in Seyfried P., Kunzmann H., McKeown P. and Weck M. (eds) *Progress in Precision Engineering* Springer-Verlag, Berlin.

7 ACTUATORS AND SENSORS FOR CONTROLLED DISPLACEMENTS

This chapter presents a variety of actuator mechanisms that are capable of operating at the nanometre level. Each mechanism is discussed in terms of its relative merits such as dynamic response, repeatability, size, linearity and cost. The actuator principles discussed in the main section are, in order; piezoelectric, electrostrictive, mechanical micrometers, friction drives, magnetostriction, magnetoelasticity, shape memory alloys and bimetallic strips, electromagnetic, electrostatic, hydraulic and the Poisson's ratio drive. For convenient reference, there follows a much briefer summary of them to clarify individual advantages and to highlight the reasons for choosing between different mechanisms. Finally some comments about the relative performance of micro-displacement sensor technologies are given.

Introduction

This chapter briefly reviews some of the techniques that may be used for primary displacement transducers that are both continuous and can be operated at the nanometre level. We shall concentrate on actuation devices since information on sensors is more readily available elsewhere. 'Primary actuator' implies that the device is not being driven by another translation mechanism, lever or gearing system or some other form of coupling. In practice it is very difficult to adhere to this definition. For example, an electromagnetic actuator operates by generating a force. This must, in turn, be transformed into a displacement by pushing against a spring which, through its stiffness, will determine the displacement. Thus the actuator inherently incorporates a lever mechanism. For similar reasons there are difficulties in outlining the limitations of each individual technique so that they might be compared. Taking the above example, the electrical drive can be switched almost instantaneously and even the magnetic field changes very rapidly. This response speed cannot be exploited in practical actuators because of the inertia of its moving elements and of the system being driven. Such inertia relates strongly to the application but little to the operational principle. Thus we indicate values for properties such as response by comparing the highest performance of *well developed* variations of each technology (for the response of an electromagnetic device, Hi-Fi headset and loudspeaker design). Although not exactly a scientific approach, and until more research work is carried out in this field, we feel that the values that have been chosen are representative of what may be possible with commercially available technology.

When specifying an actuator for applications in nanotechnology, the decision is usually based upon the three following criteria:

1. Stiffness, which determines both the dynamic response and the magnitude of the transmissible forces.

2. Positional accuracy, which limits repeatability and is often enhanced by feedback control. This usually, although not always, results in an increase in the cost of such a device.

3. Range, which, at a resolution of one nanometre, tends to be rather restricted in terms of normal engineering applications. Generally, costs can be considered roughly proportional to the range/accuracy ratio for a given resolution. A value of 100 is often easily achievable whereas 10,000 is rather expensive and 10^7 is at the limit of present capability, Hocken and Justice, 1976. Additionally, there are often physical restrictions on the range of some actuators such as magnetic saturation, yield stress etc.

The following discussions present, and compare in the light of these criteria, a range of technologies and mechanisms that have the potential for applications in smooth, continuous actuators of sub-micrometre or even sub-nanometre resolution. For quick reference, Section 7.2 reiterates some of the major points briefly. Finally, for completeness, the most common forms of nanometric displacement sensors that might be used with feedback controlers are briefly reviewed.

7.1 Types of actuator

We have found no completely satisfactory order in which to place the actuation principles here. Alphabetical listing, perhaps the easiest for quick reference, tends to separate close relatives and make comparison more difficult. The chosen order is roughly that of present popularity, with the exception that closely related methods are placed together after their popular member.

7.1.1 Piezoelectric drives

Of the twenty-one crystal classes that do not have a centre of symmetry, twenty can experience a dimensional change upon application of an electrical potential gradient (electric field). Such materials are known as piezoelectric. The lack of a centre of symmetry is a necessary precursor for this phenomenon. Ten of these classes will also generate a surface charge upon heating and these are known as pyroelectric materials. Although *all pyroelectrics are piezoelectric*, the reverse is not always true: there are fundamentally different reasons for the occurrence of these two effects. In a pyroelectric material, the net dipole is altered by a change in the magnitude of polarization when it is heated. In piezoelectric materials that are not pyroelectric, the dipoles, which are invariably altered by heating, are so arranged that they compensate each other, resulting in no net change in dipole moment. Upon application of a stress some dipole moments in specific directions are increased more than the others and a net dipole change is induced. Additionally, some materials can be polarized by the application of a sufficiently strong electric field. By analogy with ferromagnetism these are called ferroelectric although they do not necessarily contain iron. *All ferroelectrics are both piezoelectric and pyroelectric.* All the commonly used actuator materials are ferroelectrics because, like magnetic materials, they can be easily and permanently polarized (commonly referred to as poling) at or near to their Curie temperature. Thus they can first be sintered to a desired shape as a ceramic and then polarized in a desired direction. Poling does not provide perfect polarization in polycrystalline materials but results in the creation of a large number of domains in which there is an overall polarization direction. If one imagines an ideal material in which there are N parallel molecules of length δ_x having a charge of $+q$ and $-q$ at each end, then the magnitude of the total polarization is given by

$$P_x = Nq\delta_x \tag{7.1}$$

The only way to change this value is to physically squeeze the poles together by the application of an applied stress or to displace the charges with an applied field. Under uniaxial conditions the total dielectric displacement, D_x, is taken as a linear combination of the two phenomena

$$D_x = \varepsilon E_x + dT_x \tag{7.2}$$

where ε is the materials permittivity at constant stress (typically in units of F m^{-1}), d is the piezoelectric strain constant (C N^{-1} or m V^{-1}), T_x is the applied normal stress (N m^{-2}) and E_x is the field strength (V m^{-1}). In fact this treatment is very much an oversimplification and in practice the effects are heavily anisotropic so that a more complete description requires tensor analysis. Excellent introductory reviews can be found in the papers of Jaffe and Berlincourt, 1965 and Gallego-Juarez, 1989. Conversely, the strain, S_x, for a given applied field or stress is given by

$$S_x = d_t E_x + s T_x \qquad\qquad (7.3)$$

where d_t is the transposed piezoelectric strain matrix (m V^{-1}) and s is the elastic compliance at constant applied field (m^2 N^{-1}).

Values for these constants and other inter-related properties for a wide range of materials are given by Jaffe and Berlincourt, 1965. It should be stressed that not only are most of these materials pyroelectric and thus very sensitive to temperature but there is an applied field above which they will lose polarity. For the very popular material lead zirconate titanate (PZT) at 25 °C this field is in the region 0.7-1.0 kV mm^{-1}. Note, in passing, that PZT is really an abbreviation for a specific material although it is sometimes used more as if it stood for 'piezoelectric transducer'.

The simple analysis indicates that, in the absence of an applied stress, there will be a net displaced charge required to obtain a proportionate strain. It is usual to use voltage drives for this type of actuator and, as such, the actual displacement will depend upon the size, distribution and stability of the electric domains within these polycrystalline ceramics as well as other influences such as homogeneity of the material. Hence, the relationship between applied voltage and displacement usually exhibits considerable hysteresis that can only be corrected by feedback control or a charge drive. The latter approach was first proposed by Newcomb and Flinn, 1982, who suggested using a charge amplifier as the driving element. This was further simplified by Kaizuka and Siu, 1988, Kaizuka, 1989, by inserting a compensating capacitor in series with the piezoelectric and using a simple voltage drive. Using this technique the percentage hysteresis could be reduced by a factor of up to five with, however, a similar reduction in the sensitivity. More importantly, creep is also reduced in similar proportion. Creep is caused by the gradual changing of domains and will quite commonly be in the region of 30% over a period of tens of minutes, Jaffe et al, 1971. This precludes using piezoelectric devices for metrological application in open loop operation, Drake et al., 1986, van de Leemput et al., 1991. The magnitudes of these effects are very difficult to assess and control and will vary from material to material and even from batch to batch. Another undesirable property of these materials is that they display cross-coupling. Not only can this cause a lateral expansion of approximate magnitude 0.25-0.65 the axial one, but often it leads to a bending which can result in angular deflections that are a significant part of the motion in miniature actuators. With good design cross-coupling often causes no more than a few percent of the total motion and some modern tubular actuators may have parasitic tilts of less than one arcsecond per micrometre of traverse.

To overcome the nonlinearities associated with piezo-actuators, closed loop control strategies have been developed and devices are now commercially available. Using capacitive feedback techniques Hicks et al., 1984, were able to control Fabry-Perot etalons (parallel flats for use in optical spectrometry) at a dynamic gap separation maintained constant at the 1 pm $Hz^{-1/2}$ level and commercial devices are now available with a range/resolution ratio of 10,000:1.* The feedback transducers for these applications include capacitance displacement sensors and direct strain gauging. Generally, a displacement of approximately 10 μm is obtained at an applied potential of 1000 V. Since they are strain generating devices, it appears possible to increase their range by making them longer. However the strain depends upon the field strength and this decreases for a given maximum drive voltage as the length is increased. In practice the displacement to voltage sensitivity is almost independent of scale. Consequently, they are more commonly applied to small scale manipulators of the order of a few millimetres or less in dimension. Direct piezoelectric drives have high stiffness since they are essentially solid load bearing rods. An axial load, W, will cause a displacement of WL/EA, where L is the length of the actuator and E and A are the elastic modulus and the cross-sectional area perpendicular to the load. Thus the maximum stiffness is given by a short length and large area. For PZT ceramics E will be of the order of 100 GPa, so a 1 mm long, 10 mm diameter device has an axial stiffness of the order of 8 GN m^{-1}. Couplings will almost certainly dominate the overall stiffness of real systems. One of the major features of such devices is their high speed of response. It is limited only by the time for a stress wave to pass along the material, which depends on the speed of sound in the material and is inversely proportional to the device length. For a PZT device, the speed of sound of the material in the polar axis is sufficiently high to enable operation at gigahertz frequencies. Not surprisingly, a major use is for ultrasonic transducers. Small devices are much used for scanning tunnelling microscopes (and related systems) where the tip is controlled by three orthogonal piezoelectric actuators which can collectively occupy less than a 2 mm cube, Smith and Binnig, 1986.

The apparent limitation on the displacement available at reasonable drive voltage can be overcome by either stacking many actuators in series or creating a bi- or multi-laminar element analogous to the bimetallic strip, van Randeraat and Setterington, 1974. Three such arrangements are shown in Figure 7.1. As the number of actuators increases both the stiffness and the dynamic performance of these devices will reduce. The bimorph of Figure 7.1b can be further enhanced by arranging the poling such that the actuator naturally bends in an 'S' shape. This relieves the flexures from restriction due to the end clamps. Using such a method Matey et al., 1987, obtained displacements of 120 μm simultaneously in all three Cartesian axes with the resolution limited by electronic and mechanical noise. There is evidence that the hysteresis will be higher in laminar designs due to creep in the bonds. This effect which

---

* Queensgate Instruments Ltd, Ascot, UK, Lambda Photometrics, Harpenden, UK.

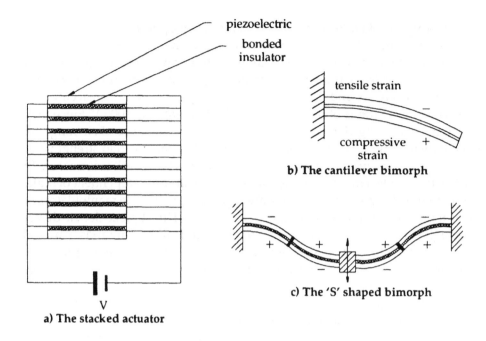

piezoelectric

bonded
insulator

tensile strain

compressive   +
strain

b) The cantilever bimorph

c) The 'S' shaped bimorph

V

a) The stacked actuator

Figure 7.1   Geometries for long range piezoelectric actuators

theoretically reduces with temperature has been observed by, Blackford *et al.*, 1987, and Tritt *et al.*, 1987. The lowest natural frequency of the displacement stage presented in these papers was 190 Hz.

For even longer range applications a device commonly called an 'inchworm*' has been developed. As the name suggests, this motion is achieved in a manner similar to that used by some caterpillars. A piezoelectric actuator is mounted onto either a slideway or an optical flat having an electrically conducting surface. Attached to the base of the actuator are two or more conducting pads separated from contact with the slideway by a thin layer of insulating material. A piezoelectric ceramic is then used as the body connecting the pads. The inchworm is actuated by first electrostatically clamping one pad to the slideway and then extending the piezoelectric material. The free end is then clamped and the fixed end is released. By then contracting the actuator and repeating these steps, it is possible to obtain unlimited motion. This type of device was first proposed by Joyce and Wilson, 1969, became famous as the coarse feed in the first scanning tunnelling microscope of Binnig and Rohrer, 1982, and was subsequently investigated by many commercial organizations. Although problems with their stability

---

*   'inchworm' is a tradename of Burleigh Instruments Inc., NY 14453, USA.

delayed wide application of commercial devices until the late 1980s, Mamin *et al.*, 1985, actuators are now available with a traverse of up to 100 mm at individual step sizes of 2 nm or less. In modern actuators the electrostatic clamps have now been replaced by piezoelectric grips. A less accurate but still interesting extension of this idea is the piezoelectric motor that operates on similar principles. There are now versions that operate by creating a circulating ripple, Tojo and Sugihara, 1985, Hatsuzawa *et al.*, 1986 rather like a self excited harmonic drive.

7.1.2 Electrostrictive

Electrostriction is superficially similar to the piezoelectric effect but in this instance strain can be conceptualized as the deformation due to a field gradient but not a direct polar strain. Although electrostrictive ceramics were first discovered in 1954, they have only recently become commercially available, Isupov *et al.*, 1989. A combination of lead magnesium niobate and lead scandium niobate is now known to exhibit a giant electrostrictive effect giving an extension of approximately 0.7 nm V^{-1}. More commonly used is a ceramic made up from 90% lead magnesium niobate and 10% lead titanate, Uchino *et al.*, 1980 and 1981. Polar strain is not the primary mechanism responsible for distortion and individual micro-domains can be considered independent in their action. Electrostriction may be better considered at this moment in time to be independent of the piezoelectric effect. Simplistically, it may be envisaged as the result of rotation of the polar domains within the ceramic. When an electric field is applied transverse to the polar direction it induces rotation of the domains resulting in behaviour characteristic of electrostatic forces in that the extension as a function of the applied field appears to follow a square law at lower values. The characteristic then becomes more linear until saturation is reached at around 10 kV cm^{-1}, Uchino *et al.*, 1981. Interestingly, due to similarities in the actuation mechanism, the strain/applied field shape characteristic for this material is of similar form to that for one branch of the magnetostrictive curve shown in Figure 7.3.

The properties of electrostricitve materials relevant to their use as displacement actuators are remarkably similar to those for piezoelectric materials, Eyraud *et al.*, 1988. Most researchers operate electrostrictive devices at a bias to avoid the strongly nonlinear region. The nonlinearity is not a real restriction if modern computer control techniques are used. As with typical piezoelectric actuators, displacements of around 10 μm are achievable at applied potentials in the region 100 V to 1 kV operating at frequencies up to 20 kHz or more. Recently, stacked systems have been produced that can achieve displacements of up to 15 μm with applied voltages around 150 V. The important difference between electrostrictive and piezoelectric actuators is that the former, although slightly less stiff, has lower hysteresis in the region of a few percent or less at room temperature and a thermal expansion coefficient in the region 10^{-6} K^{-1}. There is ambiguity in the literature about the former value because of a general confusion between linearity and hysteresis. The expansion coefficient is similar in magnitude to some of the best ceramic materials making it ideal for precision actuator applications. Reducing the temperature of the actuator below 273 K will result in a rapid increase in hysteresis obviating its benefit over the cheaper piezoelectric ceramics. Clearly, more research is required to clarify the performance of this material.

7.1.3 Mechanical micrometers

Standard micrometer screws provide a low cost solution to many short to medium range drive applications. They are invariably hysteretic due to the nonlinearity of contact forces and will display backlash unless the system is well designed. Because of the intricate and precise shape of components that make up a feedscrew it is at present only economically feasible to construct them using metals. In consequence the thermal properties of feedscrews tend to be relatively poor. However, this can often be overcome by compensation, for example by using an all steel drive loop. Even with standard micrometers position can be manually controlled to much better than a micrometre if there is visual position feedback. Open loop, they can control to perhaps 1 to 2 µm. Differential micrometers, in which there are counter-moving threads, have a more restricted range but can be manually positioned to about 0.1 µm under low load conditions and to better than 0.01 µm with feedback, Jones, 1988. The open loop accuracy is mainly restricted by pitch errors in the screw. The effect of these can be reduced by mechanical averaging using a softer material for the nut, but this invariably results in loss of drive axis stiffness. Under closed loop control positional accuracies of better than 10 nm are achievable and ruling engines have been known to repeat to a small fraction of a nanometre, Rowlands, 1902, Merton, 1949, Hall and Sayce, 1952, Stanley et al., 1968.

The stiffness of screw mechanisms is difficult to generalize. All the axial load must be carried by the threads. The deflection of any part of the thread will be a hybrid of bending and shear of cantilever forms and compression of a squat annular strut. Generally only two or three turns of the thread are likely to contribute significantly to the definition of position because of the manufacturing imprecisions. If the axial load is increased, deflection of those turns will cause more to come into close contact, so the system will act as a rapidly strengthening spring. The compliance of the interface between the thread flanks may well dominate over that of the material. For both these reasons it is best to use a relatively high axial preload in systems that depend on the drive rather than on clamps to maintain positional integrity.

Notwithstanding all these caveats, micrometer drives directly provide more than adequate stiffness for many instrument applications. Note that the ball-screw geometries used on larger systems provide for higher thread preloads without excessive penalties in torsional friction, so it is not reasonable to use data on their stiffness as indicator of what might be expected from a micrometer. Response speed is also difficult to generalize but a high-speed precision drive might typically be capable of positioning a small carriage over 100 mm range in a second or two.

7.1.4 Friction drives

There are two distinct groups of friction drives. One relies on the clamping by friction in a similar way to how we use our hands on a rope while the other utilizes the inertia of objects to overcome frictional forces in a manner akin to the trick of removing a table cloth without moving the cutlery. The inchworm mechanism described in Section 7.1.1 is something of a hybrid of both types.

The first mentioned type is the one most commonly evoked by the term 'friction drive' and is a derivative of the feedscrew and rack and pinion. It consists of a polished bar that is squeezed between two rollers. As they are rotated with high precision, the bar translates at a rate very close to one circumferential length per revolution in a good design. The main advantage of this mechanism is that the contact forces between driver roller and driven bar remain relatively constant and the range of traverse is virtually unlimited. Against these are ranged most of the problems associated with feedscrew devices plus additional ones associated with rotary accuracy/resolution, performance of the bearings supporting the squeeze rollers, surface finish of the contact, alignment of drive rod axis perpendicular to the squeeze roller axes and Heathcote slip (see Chapter 5). Slip is likely to be the most difficult to reduce. It is caused by the differing relative velocities between sliding area contacts that occur in both the ball bearings supporting the squeeze rollers and at the roller/drive rod interface. The resulting energy losses cause a hysteretic torque-displacement characteristic. If the drive is subject to any variations in load, then an additional hysteresis due to the finite slip in the contact zone would be expected, Mindlin and Deresiewicz, 1953 (see also Chapter 5). A recent investigation of both ordinary ball screws, planetary roller screws and friction drives indicated hysteresis in the region 0.03, 0.01 and 0.1 to 0.05 μm respectively, Weck and Bispink, 1991. It must be stressed, however, that the drives were acting only against a very low friction aerostatic bearing and preload techniques, which would be expected to reduce these values, were not employed. They also found that the friction drive gave smaller systematic errors, could respond to smaller step sizes and produced smaller path errors at higher speeds than the others. With feedback control friction drives exhibit a resolution in the region of ten nanometres or less and appear to be a serious challenger to the feedscrew, Donaldson and Maddux, 1984, Carlisle and Shore, 1991.

There is very little open information on friction drives for ultra-high precision applications although many such drives have been designed and used to achieve long range motion at relatively high velocity, say 100 mm s^{-1}, with sub-micron resolution, Reeds *et al.*, 1985, and devices are now commercially available covering a variety of specifications. Although the drive bar itself is solid and very stiff, the overall drive stiffness, again very difficult to quantify, is likely to be quite low because there are several contact interfaces in the system. In particular, the bearings and support shafts of the drive rollers are likely to limit what can be achieved. Even quite long, slender drive bars are likely to be stiff by comparison. All the same, a stiffness of 360 N m^{-1} given by Reeds *et al.*, 1985, seems pessimistic, given the controllability of machine tools that has been achieved with them.

The other type of drive that utilizes friction as the displacement mechanism is the dynamic translation device or inertial slider. In this a moving carriage is placed onto a slideway, held in position by frictional forces at the interface. The slideway is then subjected to an impulsive force parallel to its axis that causes the carriage to slip along the slide, because of its inertia. The slideway is then allowed to relax slowly taking the carriage with it. Repetition of this cycle results in a net motion in the direction of the impulsive force. A typical example of such a device, Pohl, 1987, is outlined in Figure 7.2. The carriage rested on the

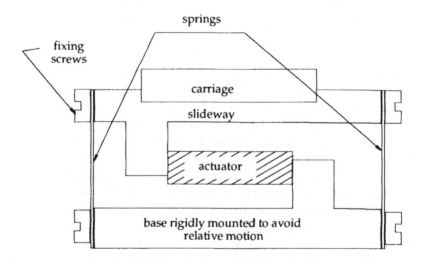

Figure 7.2   A dynamic linear spring actuator employing friction as the primary actuation mechanism, after Pohl, 1987

platform of a linear spring. A piezoelectric translator was rigidly mounted between the carriage and the base and the actuator was excited with a saw-tooth waveform to achieve the desired displacement. Using this device, it was possible to translate loads of up to 1 kg at a velocity of 0.2 mm s^{-1} with a resolution of a few nanometres. Variations of this design can be found in the papers of Blackford and Jericho, 1990, and Renner *et al.*, 1990. A similar technique was employed by Smith and Elrod, 1985, but using an electromagnetic force actuator in their case. Again, it was possible to position the carriage to within a few nanometres and the traverse speed could be controlled up to 1 mm s^{-1}. Using a combination of such devices and mounting the specimen stage on a glass flat they also achieved a two dimensional scan. The motivation behind both of the above developments was to sidestep the problems of reliability that were being encountered at that time with the inchworm type actuators. The drive stiffness is clearly low and will be almost entirely dependent on the coefficient of friction and the normal applied forces.

7.1.5 Magnetostrictive

This phenomenon is broadly similar to that of the piezoelectric effect only it involves a strain caused by the presence of a *uniform* magnetic field. This means that the displacement per unit field will increase with linear dimension making magnetostriction of interest for large scale or heavy duty actuators. As would be expected of a magnetic property, there is intrinsic hysteresis and there will be creep with time. A recently developed material of considerable interest is an alloy of terbium, dysprosium and iron (called Terfenol-D) which has giant magnetostrictive properties.

Figure 7.3 shows a plot of strain against magnetic field. When Terfenol is subject to an externally applied stress of up to around 20 MPa, a moderate field strength of 70 kA m^{-1} will result in a strain of approximately 1000 x 10^{-6}, Clark *et al.*, 1988. The strain is dependent upon the applied stress and will fall off rapidly to either side of the optimal range. Consequently, the actuator requires a preload to ensure that, for this material composition, the working stress is maintained between 8 and 20 MPa, Oswin *et al.*, 1988. The size of such an actuator is restricted primarily by the cost of the material (terbium is very expensive) and that of the field generator. A reasonable device for the control of a grinding wheel (say) might be 100 mm long and have an area of 20,000 mm^2 to give a displacement range of the order 100 μm under a load of 400 kN with a local stiffness of around 20 N nm^{-1}. This clearly has potential as a heavy duty nanoactuator under feedback control. The speed of response of such a device is restricted mainly by the inertia of the material and for the large scale applications being considered, a resonant frequency in the region of 1 kHz might be expected, Engdahl and Svensson, 1988. Yuan *et al.*, 1987, have used

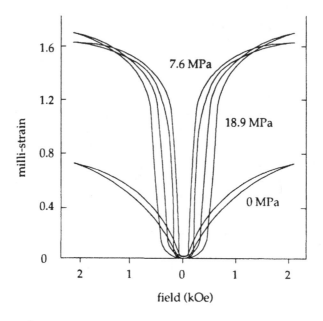

Figure 7.3   Magnetostriction in Terfenol-D (Clarke *et al.*, 1988)

this type of actuator for the closed loop control of a tool feed compensator with a 125 Hz bandwidth at 10 nm resolution. However, as the size, and thus displacement amplitude, reduces, devices having an operating frequency of up to 300 kHz are possible, Oswin *et al.*, 1988.

7.1.6 Magnetoelasticity

Frequently referred to as the ΔE effect, magnetoelasticity is related to the magnetostrictive coefficient, Squire, 1990. All magnetostrictive materials will change elastic modulus in the presence of a uniform magnetic field and it is this, in part, that contributes to the giant magnetostrictive effect. To utilize magnetoelasticity directly as a mechanical actuator the material must first be subjected to strain via a dead weight, spring or internal stresses. These will result in a distortion of the material, the magnitude of which depends on its elastic modulus. If the material is then placed within a magnetic field the elastic modulus will change and a net displacement occur. The magnitude of the change in strain depends upon both the percentage change in elastic modulus and the magnitude of the initial distortion. The maximum practical change in modulus is around 2 to 10% and the maximum strain for many materials will be in the region of 0.1%. Thus the controllable range is approximately 20 to 100 microstrain, or a change in length of 2 to 10 μm for a 100 mm long device.

A material of particular interest is Elinvar or Ni-Span-C which is a 42%Ni-Fe alloy*. This composition results in a near zero thermoelastic coefficient at room temperatures and a thermal expansion coeffient of 3×10^{-7} K^{-1}. It does, however, undergo a change in Young's modulus of approximately 0.1% in the presence of a high magnetic field. Thus it is feasible to use it as a magnetoelastic device that is dimensionally sensitive to magnetic field but is relatively immune to small temperature variations. Its main advantage over all the other types of actuator discussed is that the displacement per unit field strength will depend upon the load condition and therefore will be *tuneable*. Although the effect is generally rather small, there are materials with much greater sensitivities to the field, in particular Terfenol-D, and up to 20% strains are possible with some Ni-Fe and Co-Fe alloys, Bates, 1953. The change in elastic modulus is not linear with field strength. Its characteristic can have both positive and negative slope regions and there may be considerable hysteresis. It is best applied as part of a closed loop system. Again the stiffness and the applicability to large scale devices for such an actuator are similar to those of magnetostrictive ones. Lower cost alloys can be used in a magnetoelastic actuator.

7.1.7   Shape memory alloys and bimetallic strips

Some alloys, notably those of nickel and titanium, undergo a reversible phase transformation between martensitic and austenic structures that have very different elastic properties. The martensite is weak and can tolerate large distortions without permanent damage, while the austenite is at least five times stiffer. Suitably distorted in the martensitic phase, raising them above their austenite transition temperature will cause the distortion to disappear, with potentially large amounts of work being done on external systems. Such materials are called shape memory metals (or alloys) because they can be switched between two 'remembered' shapes. For example, the material is bent at a high temperature and then allowed to cool whereby it will return to its original shape. Upon reheating the material will return to its deformed shape. For a review of these materials see Schetky, 1979, or Golestaneh, 1984. Actuators using this effect are capable of very large displacements for a relatively small device size. A typical example would be a deflection of 20 mm for a cantilever beam of around 100 mm in length and a drive force of 300 N with a time constant of approximately 4 seconds, Yaeger, 1984. To obtain larger displacements the distorted form involves bending, so the stiffness of the drive is not as high as might be expected for a solid metallic device. It is also different in the two states and can be quite high in the austenitic phase. They require heat as the actuating mechanism, which can be a disadvantage in precision systems.

---

*   Inco Alloys International Ltd, Hereford, UK.

Heating is usually applied by passing a current through the actuator. More recently, smaller actuators have been used to align optical fibres by providing the pushing action to position them in precisely formed vee grooves. Using this technique individual fibres could be positioned to within 100 nm, Jebens *et al.*, 1989. There is also interest in using memory metal actuators for miniature robot arms, Kuribayashi, 1989.

Bimetallic strips exploit the bending caused by the differential thermal expansion of two intimately contacted thin materials as they are heated. Displacement is a direct function of temperature. They are used more often as switching elements, for example in thermostats, than as stand alone actuators. A well known example of the latter is the old vehicle direction indicator mechanism. Again, under controlled conditions, it is likely that similar range and resolution to the shape memory alloys could be achieved.

7.1.8 Electromagnetic

There are a large number of variations of electromagnetic actuation. Most involve the magnetic attraction of a spring mounted soft iron target towards a solenoid coil or an electromagnet. Such devices generate a force typically varying with the square of the width of the air-gap, so acting with normal springs gives a highly non-linear characteristic. They also involve the regular magnetization and demagnetization of soft magnets, so their repeatability with open loop control is open to doubt. They are mostly used, and excellent for, fixed actions driving hard against mechanical stops, as in electrical relays and solenoid controlled valves. They are not well suited to precise displacement control and will not be further discussed. In this section, only nominally linear devices which utilize permanent magnets in the presence in magnetic fields insufficient to induce significant demagnetization will be considered. The most common AC linear displacement actuator is the audio loudspeaker.

In precision electromagnetic devices, a saturated permanent magnet (usually SaCo or NdBFe) is surrounded by a coil (or vice versa). A force is generated between the magnet and the coil that is proportional to both the strength of the magnet and the gradient of the field from the coil perpendicular to the axis of the magnet. Thus, for a fixed geometry, the force depends directly on coil current. To obtain a controlled displacement, the force is applied to a linear spring and it is this which appears normally to limit the achievable accuracy of such devices. Using a spring manufactured from single crystal silicon, open loop displacements of up to 100 nm have been obtained with a resolution of 5 pm as measured using x-ray interferometry, Chetwynd, 1991. This type of actuator requires some form of stiffness against which to act since it is a pure force generator. It can only be considered as a displacement actuator when associated with some secondary mechanism. Because of this the available range and resolution depends critically on the geometry of the application and it is not helpful to give generalized information.

Smith and Chetwynd, 1990, show that the output force per unit input power is dependent upon both the permanent magnet material (in particular its intrinsic magnetization) and the conductivity of the input windings. The use of superconductors would be ideal, if achievable at reasonably high temperature, enabling the implementation of extremely linear, high force, low dissipation actuators. The linearity of present devices depends on the fractional demagnetization which is, in turn, closely related to the recoil permeability. It should be as low as possible for maximum linearity. Demagnetization curves for the materials NdDyFeBNb, NdDyFeB and NdFeB are shown in Figure 7.4 as curves a, b and c respectively, Parker *et al.*, 1987. Nonlinearities in the force-current characteristic occur because of demagnetization as the intrinsic magnetization follows the characteristic curve. Complete demagnetization occurs in the presence of sufficiently high field strengths (referred as the intrinsic coercivity). However, the recoil permeability will also be related to the applied field and reduces to zero at very low fields. In nanotechnological applications relatively weak fields are employed and it is the constancy of the *B-H* characteristic that limits the linearity of operation. From Figure 7.4 it can be seen that the simple NdFeB material has the flattest characteristic in the presence of a low field strength, with the introduction of dysprosium and

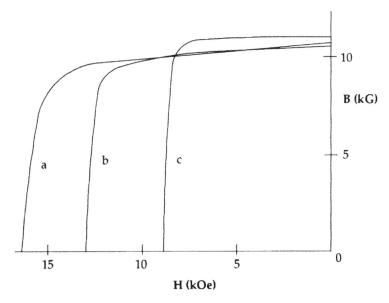

Figure 7.4 Demagnetization of some permanent magnet materials, Parker *et al.*, 1987.

niobium creating a larger slope but also considerably increasing the coercive strength. The Curie temperature (the temperature at which permanent magnetization is lost) is approximately 270 °C for these materials, although it can be increased to approximately 500 °C by the addition of cobalt, Sagawa *et al.*, 1984.

There are two different ways of utilizing electromagnetic force actuation as being considered here, moving coil and moving magnet configurations. In the first case a permanent magnet is constructed to consist of a permeable loop that is broken by a small gap in which the driven coil of winding length, L, is positioned. By passing a current, I, through this coil it will experience an axial force, F_x, given by (for a detailed discussion see Hadfield, 1962)

$$F_x = B_x IL \tag{7.4}$$

where B_x is the flux density across the gap. From considerations of continuity it can be estimated from the equation

$$B_x \approx \left(\frac{\mu_0 H_m B_m V_m}{V_g}\right)^{1/2} \tag{7.5}$$

where μ_0 is the permeability of the gap, $H_m B_m$ is the energy product for the magnetic material, and V_m and V_g are the volumes of the magnet and gap respectively.

Equation 7.5 shows that the optimum condition is obtained for a magnet material having the highest energy product and the minimum gap volume. Thus it is best to use the largest volume of magnet possible within any space restrictions. The biggest drawback with this system is that the active element, to which power must be supplied, is attached to the moving stage and energization of the winding will create a source of heat in the mechanism. One way of overcoming this problem is to use two windings, wound in opposite senses, and to operate the actuator in differential mode while maintaining a constant total input power. This should, in practice, result in a constant heat input that will eventually lead to thermal equilibrium with the temperature remaining constant during operation. The disadvantage with this method is that the extra winding is redundant in terms of the actuating force but requires additional gap space, thereby reducing the force per unit power, and adds mass to the moving stage. Because of the large mass of magnet required for this type of actuator, it is not feasible to attach the passive magnets to the moving platform.

Moving magnet designs employ a totally different geometry. In this, the permanent magnet, attached to the moving platform, is surrounded by a coil that is rigidly mounted to the base of the device. The force on the magnet is then proportional to both the magnitude of the magnetic moment and the gradient of the field along its axis of magnetization. In the case of a permanent magnet material having a constant flux density with applied field (a condition nearly satisfied by the selection of a magnet of low recoil permeability and avoiding strong fields, as discussed above) the moment is equal to the intrinsic magnetization multiplied by its volume. Additionally, under these conditions, the intrinsic magnetization becomes equal to the remanence of the magnet, B_{rem}, and the resultant force on the magnet is given by

$$F_x = B_{rem} V_m \frac{H_{x1} - H_{x2}}{x_1 - x_2} \tag{7.6}$$

where H_{x1} and H_{x2} are the values of applied field at each of the poles of the magnet which are situated at positions x_1 and x_2 (these are usually near to the ends of a permanent magnet) and F_x is the force along the x-axis, which is colinear with the vector (x_1, x_2). For a relatively short magnet in a field of gentle gradient this can be written in differential form

$$F_x = B_{rem} V_m \frac{dH_x}{dx} \tag{7.7}$$

A simple actuator can be constructed by fixing or glueing a series of disc type magnets to a linear spring device and surrounding them by a uniformly wound circular cylindrical coil. The maximum field gradient in this instance is near to each end of the coil and thus it is efficient to place magnets at both positions with their poles opposing (the gradients are equal and opposite at each end), see Figure 7.5. This type of device has the advantages that the force on the magnet is relatively insensitive to position and there is no mechanical coupling, which combination results in a near zero stiffness force actuator. A disadvantage is that the coil should be as close to the magnets as is possible to give good sensitivity. This is hard to achieve without the magnets touching the winding spool. A useful method for checking this is to electrically isolate the magnets from the spool and then check for a short circuit as an indicator of mechanical contact. The optimum geometry of the coil and considerations of both power dissipation per unit force and minimum sensitivity to small displacements of the magnet relative to the coil for such an actuator is given by Smith and Chetwynd, 1990. The conditions are not highly critical and, broadly, for small systems a coil of length equal to its outer diameter and twice its inner diameter will be satisfactory if the magnets are placed at the neck of the coil. The magnets should be fairly short compared to the length of the coil. Adjustment *in situ* is easy since the minimum displacement sensitivity occurs

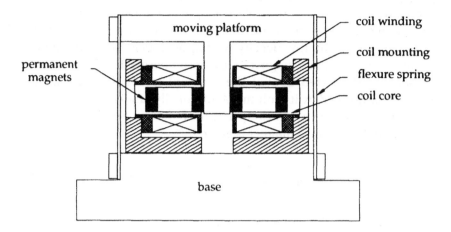

Figure 7.5   A two coil, four magnet, solenoid/magnet driven linear spring

at the point of maximum force for a fixed current. For an alternative approach, the design parameters for a coil having a uniform field gradient are presented by Pouladian-Kari *et al*, 1988.

Some idea of the potential performance of electromagnetic drives is given by the common loudspeaker, which may be one of the most efficient mechanisms, achieving displacements in the region of 500 μm over a 20 kHz bandwidth. However, it is a low stiffness form and the flexures would have to be considerably redesigned to achieve dimensional stability. The practical actuator based on the moving coil geometry is often referred to as a voice-coil linear motor, acknowledging its relationship to the loudspeaker. It has been used in applications such as computer disk drives where precise, fast indexing between tracks is required over distances of up to 50 mm. As only the coil moves it can have very low inertia. Smaller versions are used in compact disc players. Moving magnet types are applied mainly to the very precise short range movement of small, relatively stiff flexure systems. The x-ray interferometer configurations, mentioned earlier in this section and case studies discussed in Chapters 4 and 5, are typical examples. There is no doubt that easily controlled sub-nanometre level resolution is possible.

7.1.9   Electrostatic

Broadly analogous to electromagnetic transducers, electrostatically driven non-contact force actuators can be made, since any two electrodes separated by an insulating barrier will be attracted towards one another if there is an electric potential difference between them. For a rectangular, parallel plate capacitor in which flat electrodes overlap each other, the forces parallel and normal to the plates are respectively

$$F_a = \frac{\varepsilon b V^2}{2x} \tag{7.8}$$

$$F_x = \frac{\varepsilon a b V^2}{2x^2} \tag{7.9}$$

where ε is the permittivity of the gap, a is the length of the overlap and b the width of the two electrodes, x is their perpendicular separation and V is the voltage applied across them.

The equations indicate that the force is always attractive and is related non-linearly to the obvious control variables, in particular varying as the square of the potential. The reciprocal relationship to the separation may be a source of difficulty since if the electrodes move closer the force rises and a positive feedback drive to pull-down may occur. This is a positive feature for electrostatic clamping, see also Section 7.1.1. In principle the force can be changed by varying the electrical polarization in the gap, but the most common form applies a varying potential to elastically mounted electrodes to effect a change in the separation. Comparing equations 7.8 and 7.9 indicates that the force, F_a, acting parallel to the plane of the electrodes is independent of the size of overlap and, for all sensible device dimensions will be much smaller than the normal force, F_x. Hence parallel motion forms are only used at extremely small separations as is the case with micromechanical devices, Trimmer and Gabriel, 1987.

The stiffness of electrostatic devices used for displacement actuators depends essentially on that of the flexure that translates force to deflection. A constant voltage bias causes an effective increase in stiffness, opposing further separation, of the electrodes, that might be relevant to clamping applications. In the normal direction the stiffness in this sense is found by partially differentiating Equation 7.9 with respect to x. As an example, a parallel plate device having an area of 100 mm^2, a gap of 0.01 mm and operating in air at a bias of 10 V would generate a force of 0.44 mN and have an effective stiffness of 22 mN mm^{-1}. The electric field in the gap is 10^6 V m^{-1}. We see that, generally, very low forces are associated with electrostatic actuators and so they are most commonly used for ultrasonic excitation, micro-miniature devices such as submillimetre sized motors, Fan et al., 1989, Tai and Muller, 1989, and microelectronic resonance accelerometers. They may find application in resonant probe (and related) microscopy for the combined excitation and monitoring of small probes. Note the effect here of scaling: electrostatic forces become significant at small gaps, while electromagnetic ones depend directly on the volume of magnet present, so one expects the former to be the more effective only in small devices. Although the force characteristic is nonlinear, it is an exact relationship extending over many orders of magnitude, restricted mainly by the constancy of the dielectric properties of the insulating barrier. A potential disadvantage of devices having air as the separating dielectric is that its permittivity will vary with both temperature and humidity and so precise control of force requires a closely controlled environment. Conversely, electrostatic systems work well in vacuum.

7.1.10 Hydraulic

In contrast to the electrostatic devices, hydraulic drives are capable of generating very large forces per unit size of actuator. Conventional rams, controlled by spool valves, and hydraulic motors are often employed in large machines because of their excellent power to weight ratio. These considerations tend to be less commonly important in precision systems and so such drives are not especially used. We comment here, therefore, only on some unusual variants.

A very stiff short-range actuator can be made in the form of a servo-controlled hydrostatic bearing, whereby displacement is obtained by applying a differential pressure. These have been used for the fine feed mechanism in ductile grinding machines to obtain displacements of up to 100 µm with a resolution of 1 nm, Miyashita and Yoshioka, 1982. These devices tend to be very linear and have a stiffness in the region 10^8 N m^{-1}. A bonus with their use is that the same device acts as a cross-slide bearing and a feed, so that the number of system interfaces is reduced. Another approach might be to use the deflection of a pressure vessel, as if hydraulic fluid were to pressurize a very stiff balloon. Consider, for illustration, a cylindrical tube with internal diameter of 100 mm. Pressures of 300×10^5 Pa are readily achievable and controllable, so the end forces could subject the tube to a tensile force of 236 kN. If a displacement 100 µm is desired, then the tube stiffness can be up to around 2.4 GN m^{-1}. This compares with the value of 0.4 GN m^{-1} achieved by Miyashita and Yoshioka using a closed loop control strategy. This method seems not to have been tried in practice and as such it is not certain whether it would be linear (although with pressures so far below the yield stress of steel, it should be relatively well behaved). A simple tube may be unacceptable for carrying the pressures wanted and more complex behaviour is then likely.

A major problem of all hydraulic systems is the elimination, or at least reduction, of noise transmitted through the fluid from the pumps and control valves. One method for reducing this is the introduction of accumulators in the system which provide damping at the expense of operating speed. They are best suited to heavy duty systems in precise machine tools rather than to small instrument systems. The speed of response of a hydraulic actuator will be restricted by the volume of fluid required for actuation and the response of the regulator mechanisms. Most current commercial systems are only capable of operating at frequencies up to a few hertz.

7.1.11 The Poisson's ratio drive

The simple model of a pressurized cylinder introduced in the previous section does not take into account the effects of Poisson's ratio in the calculation of the axial strain. Imagine the cylinder now subjected to the same 30 MPa hydrostatic pressure on the outside wall as on the inside and with a core inside it so that there are no piston-type forces acting on the ends. Now the cylinder walls are subjected to a uniform stress of 30 MPa in the radial direction. From elasticity theory, there will be an axial strain in proportion to the radial stress and to Poisson's ratio. For a steel cylinder of length and outside diameter 100 mm and 70 mm internal diameter, the extension due to this pressure will be about 4.7 µm (assuming a Poisson's ratio of 0.33). This drive will be linear. The axial stiffness of the cylinder alone is around 8 GN m^{-1} and there would be some gearing of stiffness from the pressurization. Actuators based on this principle are being considered for applications in ultra-stiff machine tool technologies, Carlisle and Shore, 1991. Since hydraulic pressurization is used, speed of operation may be limited as noted in the previous section, although, the Poisson's ratio drive does not, in theory, require a large volume and should, in principle, be capable of operating at higher speeds than direct hydraulic drives.

A variation of the principle has been used with solids by applying a controlled squeeze around a shaft via a C-clamping arrangement to provide a desired axial motion. Using this technique, and with feedback, Jones, 1988, was able to position a linear spring mechanism to within 1 nm.

7.2 Summary of actuator performance

The detail given in Section 7.1 may sometimes obscure the points that would be used initially to select candidate technologies for a given application and so brief summaries of salient points are give in Table 7.1 and in the following brief paragraphs. Primary actuator mechanisms have been discussed in terms of scaling, range, stiffness (relevant also to dynamic response), displacement characteristics, open loop accuracy and cost. Table 7.1 provides a rough guide for comparative purposes, but for design purposes more comprehensive references must be consulted. In most categories merely relative descriptors such as 'high' or 'small' are given. Most of the terms are self-evident, but a few

ACTUATOR	SCALING	RANGE	STIFFNESS	LINEARITY	CONTINUITY	ACCURACY	COST
Piezoelectric	Small	0.1-200 μm	High	Poor	Yes	10 nm	Medium
Electrostrictive	Small	20 μm	High	Poor	Yes	10 nm	Medium
Micrometer	Large	2.5-25 mm	Medium	Moderate	No	100 nm	Low
Friction Drive	Large	2000 mm	Medium	Moderate	No	100 nm	Medium
Magnetostriction	Large	100 μm	High	Poor	Yes	Not known	Medium
Magnetoelastic	Large	1 μm	High	Poor	Yes	100 nm	Medium
Shape memory	Large	20 mm	High/Med.	Poor	Yes	0.1 mm	Low
Electromagnetic	Medium	100 μm	≈ 0	Good	Yes	0.005 nm	Medium
Electrostatic	Small	100 nm	≈ 0	Square law	Yes	< 0.1 nm	Low
Hydraulic	Large	100 μm	Very High	Good	Yes	1 nm	High
Poisson's ratio	Large	5 μm	Very High	Good	Yes	1 nm	High

Table 7.1: Generalized actuator characteristics

qualifications are useful. The scaling column indicates whether the method is likely to become more effective as sizes are increased or as they are decreased. Range generally indicates the typical maximum deflection and the presence of two figures indicates different ranges for different sub-types, not a limit on useful resolution, which is the subject of the accuracy and linearity columns. Continuity indicates whether the devices can, in principle, be free from internal mechanical interfaces that might contribute, for example, to hysteretic effects. Cost relates to the purchase of a basic actuator without any closed loop control of displacement. However, each actuator has been considered as a complete unit capable of being programmed for actuation from a computer or control voltage. In, say, piezoelectric or hydraulic actuators the ceramic element or oil may be of negligible cost compared to the stabilized high voltage or fluid pressure supply units.

The main advantages and limitations of the actuation principles are summarized below in terms of the features most likely to influence their utilization in precision drives.

Piezoelectric – These are moderately cheap, especially good for miniature systems, stiff, capable of very high response speeds and usually operate with a displacement range below 10 µm. Beam bending forms have greater range at lower stiffness. Hysteresis, creep and the high sensitivity of the coupling equations with temperature limit the achievable accuracy.

Electrostrictive – Broadly similar in performance to piezoelectric actuators, except that they are distinctly non-linear, particularly at low applied fields, but offer reduced hysteresis. Generally range is sacrificed in order to operate in the relatively linear portion of the strain/field characteristic. They have a very low thermal expansion coefficient, making them ideal for high precision applications. Lower materials processing costs and simpler electrical drives may make them cheaper than equivalent short range piezoelectric systems.

Mechanical micrometers – The large range, low cost and capability of manual operation make these an attractive option for relatively simple tasks. However, at the sub-micrometre level, they must be used with care, and usually with some form of feedback, since they exhibit backlash, temperature sensitivity and other inherent nonlinearities. Their medium stiffness is adequate for many direct drive applications.

Friction drives – Although capable of higher drive stiffness and longer range, similar limitations to those of the micrometer apply. Non-linearity depends critically on the design of the friction grip to the moving stage. With feedback or using a stick-slip type drive, large traverses are possible with resolution approaching the nanometre level.

Magnetostriction – Using some of the new giant magnetostrictive alloys that are now available, relatively large displacements can now be achieved. They are strain generators, so range depends on the size of the device and they exhibit considerable hysteresis and creep. They ideally scale to large, stiff actuators for the control of large forces with sub-millimetre range, but, at present, the alloys are prohibitively expensive for large devices.

Magnetoelasticity – Actuators employing this effect depend upon external pre-stressing but are therefore 'tuneable' to operate over different ranges using the same device. Again, the preferred scaling is for large actuators with high drive forces. Hysteresis and creep are likely to relatively large.

Shape memory alloys and bimetallic strips – These usually form low stiffness, large range actuators that depend upon bending. One phase of many memory metals is inherently quite stiff. Applications to date have been mainly to small robot arms or large conventional engineering mechanisms. They require heat to activate and phase changes may cause further heat generation, resulting in temperature gradients within the drive. Devices almost invariably have very slow dynamic response and so are used in a pseudo on-off mode.

Electromagnetic – Linear, highly characterizable actuators can be produced with saturated permanent magnet drives. The displacement mechanism is indirect, usually with an electromagnetic force actuator pushing against a linear spring. Linear range is restricted by the spring stiffness and the field uniformity. They are relatively cheap and easy to implement and can have a good dynamic response. Near zero stiffness designs allow good vibration decoupling. Unless well screened, they may be sensitive to the proximity of ferromagnetic materials.

Electrostatic – Again, the actuators generate force which is transformed to displacement by elastic mechanisms. Although non-linear, the characteristic is extremely well-defined, so they afford possibly the largest range to resolution ratio of all and performance can be accurately predicted. Forces generated are very small except at close proximity of the electrodes, favouring applications to miniature actuators. They are sensitive to the permittivity of the gap separation and limited by its dielectric strength.

Hydraulic – Hydraulic drives are most useful as high force, high stiffness actuators for the fine movement of large masses. They can be highly linear within a reasonable range. Their main disadvantage is that a hydraulic pressure generation unit is both expensive and bulky and it is hard to prevent vibrations transmitting from it to the instrumentation. Dynamic response is usually restricted by the hydraulic control.

Poisson's ratio – They have similar attributes to the hydraulic drives, but with approximately an order of magnitude less range. They may allow simpler mechanical geometries and, in principle, lower volumes of working fluid, leading to lower costs for faster dynamic responses.

7.3 Nanometric displacement sensors

There are many displacement sensor technologies that are relevant to precision applications and which are well documented in many texts, for example Neubert, 1963, Beckwith *et al.*, 1981, Bentley, 1988. Although some have theoretically infinitesimal resolution, relatively few can be applied practically to nanometre or sub-nanometre measurements without the need for expensive or complex post-processing of data. Here we shall comment briefly only upon those that can. In most cases similar devices can be used at more conventional levels, so their transduction principles are well known and need not be repeated here. Proximity probes used in scanning tip microscopes are covered in a little more detail. In addition, x-ray interferometry is examined as a case study in Chapter 4.

7.3.1 Inductive gauging

There are two variants of inductive displacement sensors that are non-contact, do not need an electrical connection to the system to be measured and, because of the low force coupling, impose little influence on the measurement. However, both require that a relatively large magnetic core is attached to the system to be measured and the mass of the core and mountings may influence smaller systems. Probably the most common is known as a Linear Variable Differential Transformer (LVDT). This exploits the mutual inductance of between two coils (a primary and secondary) coupled by a magnetic core. The degree of coupling is a direct monitor of the position of the core relative to the coils. Two similar secondary windings are placed symmetrically about the primary, which is excited typically with a few volts at a few kilohertz, Figure 7.6a. For small displacements of the core about the central position, the amplitude of the output voltage across the counterwired secondary windings will be linearly proportional to the displacement. The direction of motion of the core can be determined using phase sensitive detection techniques. The second approach, generically termed variable reluctance methods, is to monitor

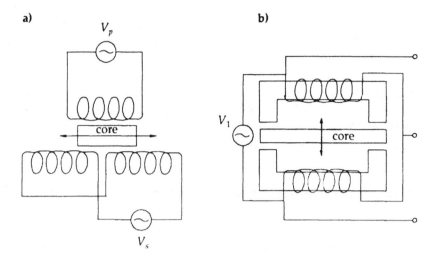

Figure 7.6   Inductive displacement transducers, a) LVDT, b) LVDI

directly the change in inductance of a coil as a highly magnetically permeable core moves in its proximity. The subsequent changes in reluctance of the magnetic circuit, and therefore the coil inductance, is again a measure of the relative displacement of the core. Simple proximity gauges are non-linear and of little use in our context. A linear, highly sensitive version is obtained by a symmetrical arrangement of a core between two coils that operate differentially. It is known as Linear Variable Differential Inductor (LVDI), Figure 7.6b, and normally incorporated as two arms of an AC bridge.

High sensitivity inductive gauges use ferrite cores for several reasons. They have high permeability (a monitor of the ratio of flux to magnetizing force). Their electrical resistivity is high (usually 100-1000 Ωcm^{-1}) so eddy current losses and consequent heat generation in the core will be negligible. Being ceramic they are relatively stable, corrosion resistant, low density, but also relatively easy to fracture or break. Both LVDT and LVDI gauges can provide discrimination to well below 0.1 nm with useful bandwidth, Whitehouse *et al.*, 1988.

7.3.2   Capacitive gauging

Electrical capacitance can also provide a non-contacting transduction method of relatively low influence. The relationships between the capacitance, C, and the charge held, q, of two parallel plate electrodes in the presence of a field of strength E depend on the permittivity of the dielectric ε between the electrodes, their area A and separation d as

$$q = CV = \frac{\varepsilon A V}{d} \qquad (7.10)$$

where $V = Ed$ is the potential difference between the electrodes. As seen in Section 7.1.9, there will be electrostatic forces attracting the plates towards each other, given by Equations 7.8 and 7.9. In most gauging configurations they are small enough to be neglected. Using Equation 7.9, a pair of 6 mm diameter parallel plate electrodes having an air gap of 10 μm will be subject to an attractive force of approximately 1 μN at 1 V.

There are many techniques for monitoring capacitance of which the AC bridge is by far the most popular for high precision displacement measurement at lower frequencies and straightforward charge amplification is used for ultrasonic measurement. Although charge amplifiers are capable of resolution equivalent to small fractions of picometres, they cannot be used at lower frequencies because of charge leakage either across the capacitor itself or in the signal conditioning electronics. Several techniques, including capacitive and piezoelectric sensing can give picometric resolution over limited bandwidths around, say, 1 MHz, Sachse and Hsu, 1979, but they are rarely usable under the static conditions of concern here. Bridge measurement techniques, using precision electronic components, can detect changes in capacitance to better than 1 part in 10^6 with a typical resolution in the region of one attofarad.

Equation 7.10 shows that it is possible to monitor relative displacement of the electrodes in a direction either perpendicular or parallel to the plane of separation. The latter requires overlapping electrodes with the change in capacitance being proportional to the increase in area. For rectangular electrodes moving in the direction of one of the principal axes, the change is proportional to displacement. Capacitance gauges of this type, comprising two concentric electrodes, with ranges of tens of millimetres and sub-micrometre resolution are commercially available. However, for operation with nanometre accuracy it becomes increasingly difficult to ensure concentricity, and therefore parallelism, between the electrodes and it is more common to monitor the change in capacitance due to a change in electrode separation. This mode is sometimes referred to as capacitance micrometry. The capacitance sensitivity, S, to change in displacement can be obtained by differentiating Equation 7.10 with respect to d to give

$$S = \frac{dC}{dd} = -\frac{\varepsilon A}{d^2} \qquad (7.11)$$

The sensitivity of capacitance micrometry increases inversely with the square of the gap. This does not, however, enable unlimited resolution because the actual capacitance value also increases inversely with gap. The area must be reduced in proportion to the separation if the nominal capacitance is to be maintained, as must generally be done to suit the electronics. Practically it becomes increasingly difficult to maintain separations below a few micrometres with electrodes of only a few square millimetres in area. Additional complications arise due to non-uniformity of the electrode surfaces and fringe field effects. At the limit it has been possible to use fine probes to detect surface proximity with gaps of only a few tens of nanometres for scanning probe microscopy applications, Matey and Blanc, 1985, Bugg and King, 1988. The effect is then almost entirely from the fringe field and linearity can be sacrificed since they are used as nulling probes for following the surface profile.

For precision capacitance micrometry it is more usual to use circular or rectangular plate electrodes of a few millimetres in dimension and separations of between 5 and 500 µm, giving typical capacitance values in the region 5 to 500 pF. For a 10 pF capacitor having a nominal separation of 50 µm, a bridge resolution at 1 aF corresponds to approximately 2.5 pm. In reality the resolution is noise limited and so depends on the signal bandwidth. The best modern practice for systems operating to DC appears to be around $1 \text{ pm Hz}^{-0.5}$. Proper bridge design provides a good linearity of output to displacement even though the capacitance itself is varying non-linearly. Errors and noise may arise with gap sensors from changes in dielectric coefficient between the electrodes, vibration transmitted along the cables connecting the electrodes, inadequate screening and parasitic tilt errors during translation. It is often convenient to use an air gap and changes in humidity or density of the air must then be avoided. If one electrode tilts with respect to the other while the mean separation is unchanged, there will be a net change of capacitance indistinguishable from one caused by a small displacement. Errors from this source are often difficult to reduce or quantify, Harb *et al.*, 1991, and may limit the traceable accuracy of real systems. Nevertheless, it is a very easy to use technique for providing sub-atomic resolutions. Another advantage is that thin film electrodes of negligible mass can be deposited onto almost any smooth surface and the capacitance simply calculated.

7.3.3   Optical transducers

The great majority of displacement measuring optical transducers use the interference of two coherent light beams, usually from a single laser source. In a typical interferometer configuration the beam is split into two with one beam reflecting from a datum surface while the other reflects from the surface to be measured. The beams returning from the reflections then combine and interfere. The intensity after interference will depend upon the difference in path lengths of the two beams, varying from a maximum to a minimum with a half wavelength change. For an ideal beam passing through ideal optics there will be a sinusoidal variation in intensity as the reflector is moved and so the phase of this sine wave may be interpreted as displacement. In practice neither the optics, laser light source nor the refractory medium through which the beams must pass will be ideal and interpolation of light fringes to better than one part in one hundred is extremely difficult, with the present state of the art around one part in one thousand, Downs, 1990, Hosoe, 1991. Some of the most severe limitations to the accuracy and resolution of optical interferometry are caused by imperfections in the optics resulting in unwanted reflections, changes in air refractivity and photodetector non-linearity. Using high quality optics in symmetric configurations and correcting for refractive index variations should allow displacement measurement with sub-nanometre accuracy. However, using simple interferometric designs measurement stabilities of the order 20 nm with a resolution of 5 nm would be more typical.

A variety of optical techniques have been developed for the profiling of smooth surfaces. Generally these rely on the phase difference between two beams reflected from similar regions on the specimen as it is scanned under the two focussed beams. Since special geometries are allowable that have small almost common paths, higher resolution is obtained than with conventional displacement interferometers. Variations of them might be applicable to very short range displacement measurement. One such profiler, Downs *et al.*, 1985, uses a birefringent lens to create a beam having two foci corresponding to the two orthogonal polarization planes derived from a helium-neon laser. The plane of the specimen surface is placed at the focus of one beam so that the reflection consists of one component from the focussed spot with a reference reflected from the larger area of the defocussed beam. The phase of each component represents that due to the surface height at the focussed spot and the integrated height over the area of the surrounding defocussed beam. Consequently, as the specimen is scanned, the resultant phase difference will vary according to the local surface profile. Being a differential measurement, this technique is relatively insensitive to vertical motion of the specimen which will only effect the lateral resolution. A sensitivity of better than 0.1 nm over a 16 nm range was obtained. A similar system developed by See *et al.*, 1988, utilizes two focussed beams separated by approximately 0.4 mm and of slightly different frequencies. The differential phase again provides information about the local surface heights and the frequency shift allows the detection of some surface optical effects. Nanometre level sensitivity was demonstrated over a 70 nm range. In both examples application is limited to optically homogeneous surfaces since phase changes at reflection look like changes in height, although in the latter case it may be possible to reduce this susceptibility. Lateral sensitivity is, in principle, limited by the wavelength of illumination but in the above examples practical difficulties limited it to between 10 and 40 μm.

Diffraction limits the performance of many optical systems and conventional metrology gratings cannot operate at the sensitivities being considered here. However, in combination they can enhance the performance of other systems. For example, by inserting diffraction gratings of 0.5 to 0.7 lines per millimetre in both the transmitted and return beams of an optical lever, Jones, 1959, 1961, was able to enhance its angular resolution to better than 10^{-10} radians (see Chapter 6).

Direct diffraction may be exploited by illuminating a fine grating, of, say, 8 or even 4 μm period, by a coherent laser source. If orders of diffraction are then combined, the phase of the resultant signal will be related to the spatial phase, and therefore position, of the grating. More complex systems that exploit the interference from two diffraction gratings illuminated using an incoherent source can also be constructed, Teimel, 1991. Presently, these systems appear to have a resolution of approximately 10 to 100 nm. They are used in the same manner as conventional incremental grating systems.

7.3.4 Proximity probes for scanning tip microscopes

Young *et al.*, 1972, were the first to exploit the fact that if a sharp, electrically conducting probe placed in near proximity to a conducting specimen with an electric potential between them, a flow of electrons (current) will occur that was strongly related to the probe to specimen separation. At a relatively large values of separation and potential difference the current could be accurately predicted from the electron emission theory of Fowler and Nordheim, 1928. More importantly, they realized that as the probe approached sufficiently closely to the surface, the phenomenon of electron tunnelling would dominate over field emission. The position of electrons can only be defined in terms of probability waves through Schrödinger's equation and there is a finite probability of them extending beyond the tip as an electron cloud having a density that decays exponentially with distance from the probe surface. If a *small* potential difference exists across the probe to surface gap, electrons can be captured and a current will flow that has an exponential dependence on the separation. The current density, J, across two perfectly flat electrodes of infinite area at an emitter spacing x_0 across which is an applied potential, V, is given by, Simmons, 1963,

$$J = \frac{e}{2\pi h(\beta x_0)^2}\left\{\overline{\varphi}\exp\left(-A\overline{\varphi}^{\,1/2}\right) - (\overline{\varphi} + eV)\exp\left(-A(\overline{\varphi} + eV)^{1/2}\right)\right\} \qquad (7.12)$$

where $A = 4\pi\beta x_0(2m_e)^{0.5}/h$, e and m_e are the electronic charge and mass, $\overline{\varphi}$ is the average potential barrier height between the electrodes, h is Planck's constant and β is a correction factor, usually close to unity. The left and right hand terms within the expression in braces describe the number of electrons flowing from the high potential to the low and from low to high respectively. Additionally as the voltage tends towards zero, the average barrier height is nearly equal to half of the average work function of the two materials. As the voltage increases, these conditions no longer apply and the situation becomes far too complicated to model using the assumptions made in the original derivation of this formula. At low voltage, Equation 7.12 reduces to

$$J = \left[\frac{(2m_e)^{\frac{1}{2}}}{x_0}\right]\left(\frac{e}{h}\right)^2 \bar{\varphi}^{\frac{1}{2}} V \exp\left(-A\bar{\varphi}^{\frac{1}{2}}\right)$$

$$= \frac{k_1 \bar{\varphi}^{\frac{1}{2}} V}{x_0} \exp\left(k_2 x_0 \bar{\varphi}^{\frac{1}{2}}\right) \tag{7.13}$$

where k_1 and k_2 are constants for a particular system.

At constant separation, the junction is ohmic with the absolute value dependent upon both the separation and the work function of the two electrodes. As the separation reduces there will be a corresponding exponential increase in current with displacement. Normally, the tunnelling current is fed into a logarithmic amplifier to give a linear overall displacement characteristic. Typically this phenomenon occurs only over a distance a few nanometres, but with four orders of magnitude change in current. Sensitivities of 2 V nm^{-1} are typical with picometre resolution at frequencies up to many kilohertz. If a constant resistance is maintained between the emitter and specimen, and if the work function (or, in reality, the barrier height) remains constant, then so too will the altitude of the probe above the surface. As discussed in Section 3.6, this can be used as a null signal to accurately scan the probe over a surface using a three dimensional piezoelectric system, see Section 3.6. This is the basic principle of the Scanning Tunnelling Microscope (STM). By monitoring the applied potential for each of the piezoelectric axes it is possible to image the surface with sub-atomic resolution. However, the accuracy of the three dimensional contour map that is derived from this type of instrument is severely limited by creep, hysteresis and temperature sensitivity. Recent attempts have been made to reduce these effects by incorporating optical measurements in the x and y axes, Barrett and Quate, 1991.

The STM operates only with conducting surfaces. It cannot be used to provide atomic resolution measurement on insulators or in a working environment with surface contamination and the formation of oxide (insulating) surface layers. Binnig and his co-workers solved this problem by the invention of the Atomic Force Microscope (AFM), Binnig et al., 1986.

In the AFM a fine probe is attached to a small cantilever beam which bends due to van der Waals and Lennard-Jones interatomic forces as the probe is brought into proximity with the surface. The deflection of the beam was originally monitored by a tunnelling probe, the signal from which was used to servo the complete cantilever and probe to maintain a constant force (that is, constant deflection). The majority of subsequent designs have used optical methods for monitoring the deflection of the cantilever beam. In essence, the AFM is closely related to conventional stylus techniques, but with forces in the region of nanonewtons rather than millinewtons. Atomic resolution images have been achieved using this technique, but only with much more difficulty than for STM and, in general, the AFM is considered to have a lower lateral resolution (the question is still not fully resolved). AFM imaging has again been extensively developed and commercial systems are now available. In addition to showing the geometry of insulators such as cleaved mica and polymers, the AFM has been used to image in real time biological processes such as the clotting of a protein, Drake *et al.*, 1989.

Contemporary with the development of the AFM was the utilization of the capacitance between the probe and surface as a proximity sensor. This resulted in the development of the Scanning Capacitance Microscope, Matey and Blanc, 1985, Bugg and King, 1988, The probe is either controlled to maintain a constant dielectric value for profiling directly or vertically modulated and servoed to maintain sensitivity between the tip and specimen for the examination of insulating layers and their associated electronic properties.

A whole range of other scanning tip technologies were developed during the 1980s. One uses resonant phenomena as a monitor of surface proximity, where the surface forces alter the effective damping. A standard AFM cantilever is excited into resonance by modulating the probe drive signal and the amplitude of the resultant oscillation used as a measure of the proximity of the surface, Martin *et al.*, 1987. This has the advantage that insulating surfaces can be detected at a probe stand-off of 15 nm or more. Another technique utilizes a micro-thermocouple through which the conduction of heat between the tip and specimen is measured to maintain a constant separation. A lateral resolution of 100 nm and vertical resolution of 3 nm has been achieved, Williams and Wickramasinghe, 1986.

Some probe microscopes have been developed explicitly for the imaging of soft surfaces. One is the Scanning Ion Conductance Microscope (SICM), Hansma *et al.*, 1989, Prater *et al.*, 1991. In this, a micropipette filled with an electrolyte is immersed in a reservoir of electrolyte and advanced towards the surface to be examined. As the space between probe and sample is decreased, so too is the gap through which ions can flow. The conductance between the pipette and electrolyte is monitored and used as the surface proximity signal for scanning. The surface of a micropore filter has been mapped by this means.

Optical tips are also suitable for soft materials and several variants that exploit evanescent wave phenomena have been proposed. For example, one probe consists of a fine lens of sub-micrometre diameter at the end of an optical fibre. Upon excitation by a photon beam, this will create a bright spot due to near field propagation. If it is brought into proximity with a surface, there will be a reflection in which the intensity varies rapidly with separation. Again this can be used to maintain a constant surface altitude to investigate the optical properties of the surface. An increase in the intensity of reflection due to a localized increase in the density of electron charge (plasmons) has been observed and this recent discovery is now being actively investigated, Pohl *et al.*, 1984, 1988.

7.3.5 Eddy current sensors

The eddy current probe is another electromagnetic technique for the measurement of fine displacements, but it does not require any connection or attachment to the system being measured. It detects the proximity and movement of a surface that must be conducting, but need not be, indeed is better not, ferromagnetic. A coil in the probe is energized at high frequency and induces eddy currents in the conductive surface. The mutual interaction of the eddy current and that in the coil causes its reluctance to vary with the relative displacement between probe and surface. To reduce eddy current losses in the transducer at high frequency, the probe consists of a ferrite core that is surrounded by a solenoid coil.

One advantage of this sensor is that, for high excitation frequencies, the eddy current losses will be confined very close to the measurement surface (the skin effect) and will therefore generally be independent of the thickness of the conductor. Sensitivity can be increased by an increase in the excitation frequency and the use of tuned oscillator electronics. A potential drawback is the need to maintain constant cable lengths and probably, therefore, large measurement loops. Zinke and Jacovelli, 1965, used eddy current sensing to monitor the displacement of a thin nickel foil over a range of 20 nm with a resolution of better than 30 pm, limited mainly by electronic noise. There are now commercially available devices that will operate over displacement ranges of up to 2 mm with a bandwidth ranging from DC to tens of kilohertz. Difficulties with stability and calibration have generally precluded them from ultra-precision applications at the nanometre level.

7.3.6 Strain gauges

If a material is subject to a uniform strain it is in theory simple to integrate over its length to determine the relative motion between two points. Resistive strain gauges provide a convenient, and common, way of realizing this idea. A simple gauge can be produced by attaching a thin (preferably less than 20 μm diameter) wire to each of the two measurement points. Any relative motion of the points will strain the wire. The change in area and length of the wire will alter its resistance and there will be a further change as the lattice distortion directly affects the electronic properties (the piezoresistive effect). Strain generally causes a larger, but nearly proportional, fractional change in resistance. The constant relating the fractional change to strain is known as the gauge factor. For metallic gauges this is related to the Poisson's ratio and the electronic state of the metal and is often approximately 2 (for example with nickel-copper or nickel-chromium) and around 5 for platinum or platinum-irridium alloys. Because fractional changes in resistance are easy to measure by using a bridge circuit, a high initial resistance will result in a greater sensitivity. This is obviously aided by winding many coils of wire between the two measurement points (providing there is still longitudinal strain in it) or bonding a meandering metal foil between two pieces of paper or to an epoxy backing strip which can subsequently be glued onto a specimen subject to strain.

Because the availability of conduction electrons in a semiconductor is strongly influenced by lattice distortion, semiconductor strain gauges typically have a gauge factor nearly two orders of magnitude higher than their metal counterparts. Unfortunately, it is often difficult to exploit the extra sensitivity that this implies because they are inherently non-linear, very temperature sensitive and rather fragile. The non-linearity coupled with the high gauge factor could also, under certain circumstances, accentuate the non-linearity of many bridge-based measuring circuits.

Typically a precise measuring system will be capable of discriminating strains of the order 10^{-6}. (Very high sensitivity strain measurement tends to use indirect methods based on optical or capacitive techniques.) If, for example, the gauge is bonded to a piezoelectric ceramic of length of the order one millimetre then it should be possible to measure at the nanometre level. However, gauges are usually more than a few millimetres in dimension and the various uncertainties in temperature, backing material and bonding stability at present limit the resolution to approximately 10 nm with conventional transducers.

A new and expanding role for semiconductor gauges may be arising through their integration within electro-micro-mechanical systems and in miniature monolithic sensors and actuators. They may well show improved performance in such applications because they are incorporated by direct doping.

References

Barrett R.C. and Quate C.F., 1991, Optical scan-correction system applied to atomic force microscopy, *Rev. Sci. Instrum.*, **62**(6), 1393-1399.

Bates L. F., 1963, *Modern Magnetism*, Cambridge University Press, Chapter XI.

Beckwith T.G., Buck N.L. and Marangoni R.D., 1981, *Mechanical Measurements*, 3rd ed., Addison-Wesley, Massachusetts.

Bentley J.P., 1988, *Principles of Measurement Systems*, 2nd ed., Longman Scientific, Essex.

Binnig G. and Rohrer H., 1982, Scanning tunneling microscopy, *Helv. Phys. Acta*, **55**, 726-735

Binnig G., Quate C. F. and Gerber Ch., 1986, Atomic force microscope, *Phys. Rev. Letts.*, **56**(9), 930-933.

Binnig, G., Gerber, Ch., Stoll, E., Albrecht, T.R., and Quate C.F., 1987, Atomic resolution with the atomic force microscope. *Surface Science*, **189/190**, 1-6

Blackford B. L., Dahn D. C. and Jericho M. H., 1987, High-stability bimorph scanning tunneling microscope, *Rev. Sci. Instrum.*, **58**(8), 1343-1348.

Blackford B.L. and Jericho M.H., 1990, Simple two dimensional piezoelectric micropositioner for scanning tunneling microscope, *Rev. Sci. Instrum.*, **61**(1), 182-184

Bugg C.D. and King P.J., 1988, Scanning capacitance microscopy, *J. Phys. E; Sci. Instrum.*, **21**, 147-151

Carlisle K. and Shore P., 1991, Experiences in the development of ultrastiff CNC aspheric generating machine tools for ductile regime grinding of brittle materials, in *Progress in Precision Engineering*, ed. Seyfried P., Kunzmann H., McKeown P. and Weck M., Springer-Verlag, Berlin.

Chetwynd D.G., 1991, Linear translation mechanisms for nanotechnology applications, *Measurement + Control*, **24**, 52-55.

Clark A. E., Teter J. P. and McMasters O. D., 1988, Magnetostriction "jumps" in twinned Tb0.3Dy0.7Fe1.9, *J. Appl. Phys.*, **63**(8), 3910-3912.

Donaldson R.R. and Maddux A.S., 1984, Design of a high-performance slide and drive system for a small precision machining research lathe, *Annals CIRP*, **33**(1), 243-248

Downs M.J., McGivern W.H. and Ferguson H.J., 1985, Optical system for measuring the profiles of super-smooth surfaces, *Precision Engineering*, **7**(4), 211-215.

Downs M.J., 1990, A proposed design for an optical interferometer with sub-nanometric resolution, *Nanotechnology*, **1**(1), 27-30.

Drake B., Sonnenfeld R., Schneir J., Hansma P. K., Slough G. and Coleman R. V., 1986, Tunnelling microscope for operation in air or fluids, *Rev. Sci. Instrum.*, **53**(3), 441-445

Drake B., Prater C.B., Weisenhorn A.L., Gould S.A.C., Albrecht T.R., Quate C.F., Cannell D.S., Hansma H.G., and Hansma P.K., 1989, Imaging crystals, polymers and processes in water with the atomic force microscope. *Science*, **243**, 1586-1589

Engdahl G. and Svensson L., 1988, 'Simulation of the magnetostrictive performance of Terfenol-D in mechanical devices', *J. Appl. Phys.*, **63**(8), 3924-3926.

Eyraud L., Eyraud P., Gonnard P. and Troccaz M., 1988, 'Materiaux electrostrictifs pour actuateurs', *Revue Phys. Appl.*, **23**, 879-889.

Fan L., Tai Y. and Muller R. S., 1989, 'IC-processed electrostatic micromotors', *Sensors and Actuators*, **20**, 41-47.

Fowler R.H. and Nordheim L., 1928, Electron emission in intense electric fields, *Proc. Roy. Soc. Lond.*, **A229**, 173-181.

Gallego-Juarez J.A., 1989, 'Piezoelectric ceramics and ultrasonic transducers', *J. Phys. E: Sci. Instrum.*, **22**, 804-816.

Golestaneh A. A., 1984, 'Shape-memory phenomena', *Physics Today*, **37**(4), 62-70.

Hadfield D., 1962, *Permanent Magnets and Magnetism*, J. Wiley and Sons, London, Chapters 2, 6 & 7.

Hall R. G. N. and Sayce L. A., 1952, On the production of diffraction gratings. II: The generation of helical rulings and the preparation of plain gratings therefrom, *Proc. Roy. Soc. Lond.*, **A215**, 536-550.

Hansma, P.K., Drake, B., Marti, O., Gould, S.A.C., and Prater, C.B., 1989, The scanning ion conductance microscope. *Science*, **243**, 641-643

Harb S., Chetwynd D.G. and Smith S.T., 1991, Application and performance of capacitance micrometry as a super precision transfer standard, *Proceedings of ASPE*, Santa Fe, October 13-18, 17-20.

Hatsuzawa T., Tanimura Y., Yamada H. and Toyoda K., 1986, Piezodriven spindle for a specimen holder in the vacuum chamber of a scanning electron microscope, *Rev. Sci. Instrum.*, **57**(12), 3110-3113

Hatsuzawa H., Toyoda K. and Tanimura Y., 1986, Speed control characteristics and digital servo system of a circular traveling wave motor, *Rev. Sci. Instrum.*, **57**(11), 2886-2890

Hicks T.R, Reay N.K., and Atherton P.D., 1984, The application of capacitance micrometry to the control of Fabry-Perot etalons, *J. Phys. E.: Sci. Instrum.*, **63**(2), 49-55.

Hocken R. and Justice B., 1976, *Dimensional stability*, NASA report NAS8-28662.

Hosoe S., 1991, Laser interferometric system for displacement measurement with high precision, *Nanotechnology*, **2**, 88-95.

Isupov V.A., Smirnova E.P., Yushin N.K. and Sotnikov A.V., 1989, Electrostrictor ceramics: Physics, materials applications, *Ferroelectrics*, **95**, 179-183.

Jaffe H. and Berlincourt D. A., 1965, Piezoelectric transducer materials, *Proc. IEEE*, **53**(10), 1372-1386

Jaffe B., Cook W. R. and Jaffe H., 1971, *Piezoelectric Ceramics*, Academic Press, London, pages 83-85.

Jebens R., Trimmer W. and Walker J., 1989, Micro actuators for aligning optical fibers, Proc. Second IEEE Workshop on *Micro-Electro Mechanical Systems*, Salt Lake City, 35-39

Jones R.V. and Richards J.C.S., 1959, Recording optical lever, *J. Sci. Instrum.*, **36**, 90-94.

Jones R.V. and Richards J.C.S., 1961, Some developments and applications of the optical lever, *J. Sci. Instrum.*, **38**, 37-45.

Jones R. V., 1988, *Instruments and Experiences; Papers on Measurement and Instrument design*, J. Wiley and Sons, NY, paper XVIII, 219-247 and paper XVII, 203-204.

Joyce G. C. and Wilson G. C., 1969, Micro-step motor, *J. Phys. E; Sci Instrum.*, **2**(2), 661-663

Kaizuka H. and Siu B., 1988, A simple way to reduce hysteresis and creep when using piezoelectric actuators, *Japan J. Appl. Phys.*, **27**(5), L773-L776

Kaizuka H., 1989, Application of capacitor insertion method to scanning tunnelling microscope, *Rev. Sci. Instrum.*, **60**(10), 3119-3122

Kuribayashi K., 1989, Millimeter-sized joint actuator using a shape memory alloy, *Sensors and Actuators*, **20**, 57-64

van de Leemput L.E.C., Rongen P.H.H., Timmerman B.H. and van Kempen H., 1991, Calibration and characterization of piezoelectric elements as used in scanning tunneling microscopy, *Rev. Sci. Instrum.*, **62**(4), 989-992.

Mamin H. J., Abraham D. W., Ganz E. and Clarke J., Two-dimensional, remote micropositioner for a scanning tunneling microscope, *Rev. Sci. Instrum.*, **56**(11), 2168-2170.

Martin Y., Williams C.C. and Wickramasinghe H. K., 1987, Atomic force microscope force mapping and profiling on a 10 nm scale. *J. Appl. Phys.*, **61**(10), 4723-4729

Matey J.R. and Blanc J., 1985, Scanning capacitance microscopy, *J. Appl. Phys.*, **57**(5), 1437-1444

Matey J. R., Crandall R. S. and Bryki B., 1987, Bimoph-driven x-y-z translation stage for scanned image microscopy, *Rev. Sci. Instrum.*, **58**(4), 567-570

Merton T., 1950, On the reproduction and ruling of diffraction gratings, *Proc. Roy. Soc. Lond.*, **A201**, 187-191.

Mindlin R.D. and Deresiewicz H., 1953, Elastic spheres in contact under varying oblique forces, *Trans. ASME; J. Appl. Mech.*, **75**, 327-344.

Miyashita M. and Yoshioka J., 1982, Development of ultra-precision machine tools for micro-cutting of brittle materials, *Bull. Japan Soc. of Prec. Engg.*, **16**(1), 43-50.

Neubert H.K.P., 1963, *Instrument Transducers*, Oxford University Press, Lond.

Newcomb C.V. and Flinn I., 1982, Improving the linearity of piezoelectric ceramic actuators, *Electronics Letters*, **18**(11), 442-444

Oswin J. R., Edenborough R. J. and Pitman K., 1988, Rare-earth magnetostriction and low frequency sonar transducers, *IOP/IEE/IEEE day seminar on Magnetic Sensors and Amorphous Materials*, 20th April 1988, reproduced in *Aerospace Dynamics*, **24**, 9-13.

Parker S. F. H., Pollard R. J., Lord D. G. and Grundy P. J., 1987, Precipitation in NdFeB-type materials, *Trans. IEEE*, **Mag-23**(5), 2103-2105.

Pohl D.W., Lenk W. and Lanz M., 1984, Optical stethoscopy: Image recording with resolution $\lambda/20$, *Appl. Phys. Lett.*, **44**(7), 651-653.

Pohl D. W., 1987, Dynamic piezoelectric translation devices, *Rev. Sci. Instrum.*, **58**(1), 54-57.

Pohl, D.W., Fischer, U.Ch., and Durig, U.T., 1988, Scanning near field optical microscropy (SNOM). *Microscopy*, **152**(3), 853-861.

Pouladian-Kari R., Parkes D., Jones R. M. and Benson T. M., 1988, A multiple coil solenoid to provide an axial magnetic field with a near-linear gradient, *J. Phys. E: Sci. Instrum.*, **21**, 557-559.

Prater C.B., Hansma P.K., Tortonese M. and Quate C.F., 1991, Improved scanning ion-conductance microscope using microfabricated probes, *Rev. Sci. Instrum.*, **62**(11), 2634-2638.

Reeds J., Hansen S., Otto O., Carroll A. M., McCarthy D. J. and Radley J., 1985, High speed precision X-Y stage, *J. Vac. Sci. Technol.*, **B3**,(1), 112-116.

Renner Ch., Niedermann Ph., Kent A.D. and Fischer O, 1990, A vertical piezoelectric inertial slider, *Rev. Sci. Instrum.*, **61**(3), 965-967.

Rowlands H. A., 1902, *The Physical Papers of H. A. Rowlands*, The John Hopkins Press, 691-706.

Sachse W. and Hsu N.N., 1979, Ultrasonic Transducers, In *Physical Acoustics*, Chapter 14, 277-407, Academic Press, New York.

Sagawa M., Fujimmura S., Yamamoto H., Matsuura Y. and Hiraga K., 1984, Permanent magnet materials based on the rare earth-iron-boron tetragonal compounds, *Trans. IEEE*, **Mag-20**(5),1584-1589.

Schetky L. M., 1979, Shape memory alloys, *Scientific American*, 74-82.

See C.W., Appel R.K. and Somekh M.G., 1988, Scanning differential optical profilometer for simultaneous measurement of amplitude and phase variation, *Appl. Phys. Lett.*, **53**(1), 10-12.

Simmons J.G., 1963, Generalized formula for the electric tunnel effect between similar electrodes separated by a thin insulating film. *J. Appl. Phys.*, **34**(6), 1793-1803

Smith D. P. E. and Binnig G., 1986, Ultrasmall scanning tunneling microscope for use in a liquid helium storage dewar, *Rev. Sci. Instrum.*, **57**(10),2630-2631.

Smith D. P. E. and Elrod S. A., 1985, Magnetically driven micropositioners, *Rev. Sci. Instrum.*, **56**(10), 1970-1971.

Smith S. T. and Chetwynd D. G., 1990, An optimized magnet-coil force actuator and its application to linear spring mechanisms, *Proc. Instn. Mech. Engrs.*, **204**(C4), 243-253.

Squire P. T., 1990, Phenomenological model for magnetization, magnetostriction and ΔE effect in field-annealed amorphous ribbons, *J. Magnetism and Magnetic Materials*, **87**, 299-310.

Stanley V. W., Franks A. and Lindsey K., 1968, A simple ruling engine for diffraction gratings, *J. Phys. E: J. Sci. Instrum.*, **1**(2), 643-645.

Tai Y. and Muller R. S., 1989, IC-processed electrostatic micromotors, *Sensors and Actuators*, **20**, 49-55.

Teimel A., 1991, Technology and application of grating interferometers in high precision measurement, in *Progress in Precision Engineering*, ed. Seyfried P., Kunzmann H., McKeown P. and Weck M., Springer-Verlag, Berlin.

Tojo T. and Sugihara K., 1985, Piezoelectricdriven turntable with high positioning accuracy, *Bull. Japan Soc. of Prec. Engg.*, **19**(2), 135-137.

Trimmer W. and Gabriel K., 1987, Design considerations for a practical electrostatic micromotor, *Sensors and Actuators*, **11**, 189-206

Tritt T. M., Gillespie D. J., Kamm G. N. and Ehrlich A. C., 1987, Response of piezoelectric bimorphs as a function of temperature, *Rev. Sci. Instrum.*, **58**(5), 780-783

Uchino K., Cross L.E. and Nomura S., 1980, Inverse hysteresis of field induced elastic deformation in the solid solution 90 mol% $Pb(Mg1/3Nb2/3)O3$ - 10 mol% $PbTiO3$, *J. Mat. Sci.*, **15**, 2643-2646.

Uchino K., Nomura S., Cross L.E., Newham R.E. and Jang S.J., 1981, Review: Electrostrictive effect in perovskites and its transducer applications, *J. Mat. Sci.*, **16**, 569-578.

van Randeraat J. and Setterington R. E., 1974, *Piezoelectric Ceramics*, Mullard Ltd, London.

Weck M. and Bispink T., 1991, Examination of high precision slow motion feed drive systems for the sub-micrometre range, in *Progress in Precision Engineering* ed. Seyfried P., Kunzmann H., McKeown P. and Weck M., Springer-Verlag, Berlin.

Whitehouse D.J., Bowen D.K., Chetwynd D.G. and Davies S.T., 1988, Nano-calibration for stylus-based surface measurement, *J. Phys. E: Sci. Instrum.*, **21**, 46-51

Williams C.C., and Wickramasinghe H.K., 1986, Scanning thermal profiler. *Appl. Phys. Letts.*, **49**(23), 1587-1589

Yaeger J. R., 1984, A practical shape-memory electromechanical actuator, *Mechanical Engineering*, 51-55.

Young R., Ward J., and Scire F., 1972, The topographiner: An instrument for measuring surface microtopography, *Rev. Sci. Instrum.*, **43**, 999-1011.

Yuan Z., Feng Z. and Zhao W., 1987, Quick responding magnetostrictive micro-feeder, *Proc. 6th Int. Conference on Precision Engg.*, Osaka, 593-598.

Zinke O. and Jacovelli P.B., 1965, Nanocentimetre transducer, *Rev. Sci. Instrum.*, **36**(7), 914-915

Zinke O. and Jacovelli P.B., 1965, Magnetic flux sensor, *Rev. Sci. Instrum.*, **36**(7), 916-920

8 MATERIALS SELECTION IN PRECISION MECHANICAL DESIGN

The selection of a best, or at least a good, material for a design is part of every mechanical engineering project. There is much data that needs to be examined in a systematic way and various schemes for so doing have been proposed. In the particular fields of precision design and instrument engineering, the parameters chosen in these general studies may not be wholly appropriate. Also a wider range of exotic materials may become feasible when only small structures are being used. This review examines the materials property needs of precision engineering and compares graphical techniques that may be helpful in the selection process. Particular attention is given to a 'profile' of property groups (ratios and products). This is followed by some general observations about the applicability of a range of materials and some examples of the compromises inherent in materials selection.

Introduction

The problem of materials selection has been with us as long as there has been design. Virtually all books on design pay some attention to it and there are textbooks dedicated to it, for example Charles and Crane, 1989. Methods of systematizing the selection process have been proposed in academic papers, as discussed below. Since they are addressed to general audiences they tend to concentrate on the needs of well known and economically important areas of general mechanical engineering and aerospace. Our concern is with the more specialized fields of precision mechanical engineering and instrument building. Here the dominant concerns may be different, for example strength to weight ratios (critical in aerospace applications) may not matter too much but the ability to maintain stability of shape, often to a high degree, probably does. Other options may arise because material is used in small amounts for one off or small batch production. Actual material cost may therefore be somewhat less important to the total costs and so higher priority can be given to obtaining the limits of possible performance. This chapter explores methods for rapidly discovering a relatively small set of candidate materials from amongst the many available, with special reference to the needs of instrument design.

Whatever the field, it is relatively rare that one single property totally dominates a design (and if it does selection is much simplified), so the idea of property ratios is almost universally used. Actually 'ratio' is rather too narrow a term, for sometimes products of properties may be better. Hence here the expression 'groupings' will be used. In many technological fields parameter groups are used as a way of non-dimensionalizing so that models may be easily compared. Note carefully that this is not the case here; the groupings are usually expressed in natural units and can be directly compared only when expressed in identical terms. There is also a question of how to compare materials realistically, for it is only worth changing or using a more expensive alternative if there is a significantly large benefit. Graphical presentation methods that summarize general trends have been independently developed to address these issues. Inevitably some form of data tabulation underlies the graphs and so these methods are immediately attractive for use with simple computer databases and spreadsheets.

This chapter is concerned mainly with the choices to be made between existing materials rather than with the design of new ones. It starts with some historical examples of the interaction of mechanical design and materials availability and goes on to explore the basic properties of mechanical systems that are most important to precision applications. This leads to a review of property chart methods and the idea of a property profile. Some rather general comments about a range of common and more esoteric materials then leads to a few case study examples of how selection may operate in practice. There are unfortunately few readily accessible sources of information on materials for instrument design. There are plenty of reference books with property tables and more traditional materials are often discussed in older books on experimentation such as Braddick, 1966. Interesting uses of newer materials are to be discovered serendipitously within articles that describe specific instruments.

8.1   Design, materials and function

An excellent early example of the interaction of materials availability and design philosophy is found in the development of chronometers. This was used to illustrate ideas of compensation in Chapter 3 and here it is reiterated from a materials perspective. Thermal expansion of a pendulum alters the accuracy of a clock and so limits its usefulness. To overcome this, the simple pendulum rod was replaced by a serpentine arrangement of rods of different expansion coefficient connected alternately at the bottom and top only. In principle by adjusting the relative lengths of rods the 'downwards' expansion of the longer ones can be compensated by an 'upwards' expansion of the shorter. In practice it is difficult to get close to perfection, but relatively easy to do better than the simple rod of either material. An advantage of using mixed materials like this to provide thermal compensation is that, within reason, readily available materials may be used. An alternative approach is to discover or to design an alloy that has almost zero expansion so that a simple mechanical arrangement can be used. This might not be as difficult as at first it seems because often the expansion need be near zero for only a limited range of ambient temperatures. In a suitably complex alloy a phase change, with a consequent reduction of volume, may occur at near room temperature that will compensate for the normal expansion of small temperature increases. The major success in this arena was the introduction of an alloy of iron with about 36% nickel, commonly called Invar, although this is actually a trade name. After suitable heat treatment, it has an expansion coefficient of less than one tenth that of steel.

For portable chronometers the pendulum is not practical and so a spring and balance wheel, that is a classical second order system, is used as the harmonic oscillator. There are now two sources of thermally induced inaccuracy. The balance wheel will expand, which is equivalent to the pendulum expansion, and also the stiffness of the spring will vary with temperature. The compensation strategy now involves the use of bimetals, a crude composite, in which two intimately joined strips of differing expansivity will bend as the temperature changes. They are used in the circumferential arms of a split balance wheel so that as it expands radially, the arms bend inwards to redistribute the mass and maintain constant moment of inertia. In principle, the compensation could over- or under-compensate by an amount that accounts for the varying elasticity of the spring. Alternatively, Invar readily provides a balance wheel of low thermal sensitivity. Another alloy then needs to be designed that has a near zero thermoelastic coefficient around room temperature. Nickel is, again, the key to doing this and increasing to about 42% its proportion in iron alloys gives materials such as Elinvar or Ni-Span-C which have the desired property.

A third approach to avoid thermal effects that is regularly used in mechanical metrology is to control the environmental temperature. Materials with higher expansivities but other desirable properties can then be more easily used. Eighteenth and early nineteenth century length standards for both the metre and the yard were specified at 0 °C so that melting ice could be used to provide a well defined temperature. The end standard for the yard was cast in a specially specified bronze, presumably in the quest for stability. The rectangular bar was heavy and distorted under its own weight. This was one motivation for the work that defined the best support positions for minimal distortion now known as the Airy points, see Appendix A. The 1889 metre was defined as a line standard based on a platinum iridium alloy chosen for mechanical stability. For this Tresca invented the, now familiar, asymmetric X section so that the calibration lines lie in the neutral plane of a beam having a high bending stiffness for its weight. Again we see the interaction of materials design with inventive mechanical design. Incidentally, the standard metre was well specified and appears to remain very stable but the yard standard has been shrinking by about one microinch per year consistently over the last century.

The illustrations quoted so far have been of open systems where lengths may change freely. These are relatively rare. Sub-assemblies of most devices act in some sense as couplings and we return to the idea of loop structures, particularly the force loops and metrology loops explored in Chapter 3. Dimensional stability of these loops is critical to many precision instruments and machines which operate under moderate variations of forces and thermal stimulus. Along with such rules as the use of symmetry, particularly with respect to heat sources, careful materials selection will be most beneficial.

The Talystep instrument designed by R.E. Reason for Rank Taylor Hobson in the early 1960s well illustrates some of these points. It is used for measuring step heights, and fine surface finish, of thin films to sub-nanometre precisions using a stylus and LVDI displacement gauge. It traverses slowly so the only significant force in its structure is self weight, which does not cause much changing distortion. Thus reduction of thermal effects in its loops was judged the critical design goal. An asymmetric, cantilever style, design was preferred for ergonomic reasons. There are two main metrology loops. One is around the sensor head and is quite small, about 30 mm. The presence of the sensor in this loop ensures that it must contain a mixture of materials. A compensation strategy was therefore adopted, with several carefully selected materials being used around the loop for no obvious mechanical purpose. Some were unusual and their presence seems to have confused some later designers, but the details are not important here. The larger loop, with characteristic dimensions of around 300 mm, incorporates the specimen traverse and the structure carrying the sensor head. It can be dominated by a single material and aluminium was chosen. Its low density is attractive. However the key decision was that its high thermal conductivity would ensure that temperature gradients and consequent loop distortions were minimized. Recalling that the Abbe offset related effects of changes in loop shape commonly dominate those of changes in loop size, this outweighed the disadvantage of a large thermal expansion coefficient. Partly from demands for increased performance, but mainly because of improvements in the material options, a different strategy was adopted in the next generation, Lindsey *et al.* 1988. In this a minimum expansion design was based on the ultra-low expansion glass ceramic 'Zerodur'. The whole design philosophy of the instrument changed because a new material became available.

8.2 Property groupings and charts

A saving grace in the potentially massive task of examining how the different properties of materials affect the performance of designs is that we can often discover archetypal behaviours: the need to carry loads; the need to have small deflections; and so on. These can be modelled for a small number of very simple structures from which we can infer the general effect of changing material in a more complex system subject to similar design goals. For example, while the transverse deflection of a beam under load depends upon how it is mounted, the *relative* effect of a change in geometry or material will be independent of the form of mountings. Consider, for example, a beam of constant rectangular cross-section, b wide and d deep. From Chapter 2 or numerous textbooks, the deflection, δ, from an imposed load, W, varies as

$$\delta \propto \frac{WL^3}{EI} \qquad (8.1)$$

If it is horizontal, its deflection due to its self-weight varies as

$$\delta \propto \frac{\rho A g L^4}{EI} \qquad (8.2)$$

where L is the characteristic length, E the modulus of elasticity (Young's modulus), ρ density, $A = bd$ and $I = bd^3/12$ are the area and the second moment of area of the cross-section and g the gravitational acceleration. Also the maximum bending moment that the beam can carry without any yielding is

$$M_{max} = \frac{2YI}{d} \qquad (8.3)$$

where Y is the yield strength (stress) in tension, or compression if lower.

For a fixed geometry, Equation 8.2 shows deflection under gravity is minimized by choosing a material with a minimum ratio ρ/E. In practice, geometries are rarely totally fixed and different criteria may then apply. For example, in vehicle, particularly aerospace, applications a beam or plate of certain stiffness and predefined length may be wanted with minimum weight. From Equation 8.1 the stiffness is governed by

$$\lambda \propto \frac{Ebd^3}{L^3} \qquad (8.4)$$

while the weight is simply $\rho Lbdg$. Both stiffness and weight depend linearly on b and L is fixed, but d may be varied to minimize weight while maintaining the specified λ. From Equation 8.4, if other terms are fixed

$$d_\lambda \propto E^{-\frac{1}{3}} \tag{8.5}$$

and then

$$weight \propto \frac{\rho}{E^{\frac{1}{3}}} \tag{8.6}$$

Thus selecting a material with minimum $\rho/E^{\frac{1}{3}}$ is optimal. Under other conditions, such as Euler buckling or bending of rods, $\rho/E^{\frac{1}{2}}$ is the best ratio to use. By incorporating also Equation 8.3, similar arguments lead to a set of criteria for load capacity in terms of strength/density ratios.

Various attempts at summarizing information about property groupings have tended independently towards common techniques. One is the use of graphical methods to present the information. Another is the use of logarithmic scales in such diagrams. Ashby, 1989, uses them mainly as a way of dealing with power laws of the type just derived to produce sets of detailed 'property charts'. Each has two orthogonal axes representing values of two properties (or other groups) commonly found in ratio, such as E and ρ or Y and ρ. A material with known values of these properties can be allocated to a specific point on the chart. By using log scales families of parallel lines representing constant values of ratios of any powers of the variables can be drawn, making comparison against sets of criteria relatively easy.

Figure 8.1 shows a much simplified property chart. It includes a line of constant $E^{\frac{1}{2}}/\rho$ from which it is readily seen, for example, that aluminium alloys are better than steels for a minimum weight strut that must support a specified buckling load. A full chart for this property pair would contain many more materials. It would also contain lines representing all the ratios discussed so far and probably also $(E/\rho)^{\frac{1}{2}}$ which corresponds to constant longitudinal wave velocity (speed of sound in thin rods). Ashby, 1989, shows many examples and others appear elsewhere. This particular paper is recommended because it has a good description of why properties group as they do and also discusses some important properties less easily shown on the charts. However, it is much concerned with features such as selection for minimum weight or for thermal shock resistance and these are not often the most critical mechanical and thermal considerations in precision design. Note, before leaving this approach, that Ashby, 1991, develops the property chart idea to include a shape factor. This is designed to reflect that with different families of materials (metals, polymers, composites, etc.) the best cross-sectional shape for, say, a minimum

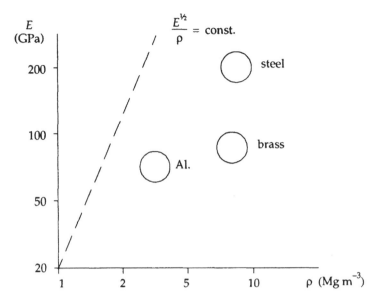

Figure 8.1 Logarithmic plot of elastic modulus against density, illustrating the concept of Ashby's property charts.

weight structure may be quite different. This is a potentially important issue, but not one that will be pursued here. A massive database has now been produced incorporating this property chart approach, Waterman and Ashby, 1991.

Chetwynd, 1987, gave different arguments for using logarithmic scales. The nature of a selection exercise is of broad scope, not of close comparisons. Some materials show considerable variation in some of their properties from sample to sample. A small improvement in performance expected from a comparison of typical values might not therefore lead to consistently better performance in practice. Also, and often of great practical importance, it is inherently expensive to make changes to an existing design. Unless we are pushing right to the technological limits, economic judgement ensures that a new material would be introduced only if it displays a considerable advantage over a traditional rival. With log scales, doublings of value occupy the same amount of space on the chart whatever the values happen to be, so the chart naturally expresses the idea of a factor of improvement. Another difference of approach is that there is no explicit concern to use as graph axes properties or groups of direct *physical* significance. Instead, a more application specific set of groupings known to occur in typical design formulae are preferred, notably those relevant to precision loop structures. This allows a rapid focussing of attention onto materials likely to be useful in this field, at the expense of distancing the study from the principles that cause the candidates to behave usefully.

SYMBOL	MATERIAL	SYMBOL	MATERIAL
Al	Aluminium	Al_2O_3	Alumina (high density)
Be	Beryllium	Bra	70/30 Brass
Bro	90/10 Bronze	CI	Cast iron
FQ	Fused quartz	Inv	Invar
MS	Mild steel	Si	Silicon single crystal
SiC	Silicon carbide, reaction bonded	SiN	Silicon nitride, reaction bonded
StSt	18/8 Stainless steel	ULE	Ultra-low expansion glass ceramics
W	Tungsten		

Table 8.1: Key to the materials shown in Figure 8.2

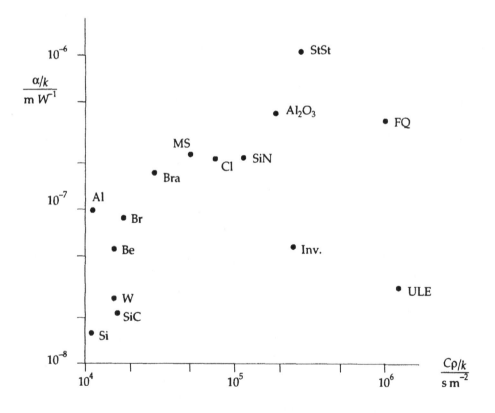

Figure 8.2   View of property map, Chetwynd, 1987

A three dimensional property space was proposed, with groupings expressed as reciprocals, if necessary, so that preferred materials plot always towards the bottom left corner of the occupied space. For precision engineering the axes suggested were α/k, ρ/E and $C\rho/k$, the latter being the inverse thermal diffusivity; α is the thermal expansion coefficient, k the thermal conductivity and C the specific heat (the values at constant pressure and constant volume are almost identical in solids). Note that the group $C\rho$ is often called the volume specific heat.

Figure 8.2 shows a view of one plane of the property space given by Chetwynd, 1987, and Table 8.1 provides a key to the abbreviations used. The reasons for the choice of axes will be explored more deeply in the next section, but, broadly, the vertical axis represents stability under steady thermal load and the horizontal axis that under transient thermal load. It indicates almost instantly that, for instance, fused quartz known for its superb temporal dimensional stability may not perform as well as expected if its thermal environment is poor. As expected, aluminium and copper alloys show good thermal performance, although several more 'difficult' materials are better. Aluminium and ULE glass ceramic appear almost diametrically opposite across the space, an interesting reflection on the different philosophies behind their use in metrology loops.

Whether either type of chart discussed or some other method is used, innovative designs may start by discovering materials that do not fit into the main clusters that occur. For example, there is a rough rule of thumb, obeyed quite well by many materials, that α is proportional to $1/E$. A property chart based on these axes rapidly shows materials that do not fit the pattern, usually because of internal phase changes, such as Invar and glass ceramics, Ashby, 1989. It must also be recognized that many selection decisions are made that cannot conveniently be shown on such charts: cost, ease of production, dimensional stability, damping and so on. Ashby does show examples of damping but succeeds in little more than demonstrating how variable it can be. The charts are probably best suited to pedagogic uses or as a planning tool for rapidly eliminating unsuitable materials and identifying potentially fruitful areas for more careful exploration.

8.3   Property profiles in precision engineering

Noting the comments at the end of the previous section and that the selection tools are always incomplete, modified procedures will be needed if materials are to be compared in a selection guide. If more than two or three groupings need to be considered, it becomes increasingly inefficient to use the charts for cross-comparisons. Chetwynd, 1989, proposed instead that the quality of one material over a wider range of groupings be summarized graphically in a property profile. It was reasoned that comparing the profiles of different materials was easier than comparing the property maps. Before exploring the structure of such profiles it is best to look in more detail at the choice of groupings that they might contain if they are to be used primarily for precision engineering applications. The two major problem areas for mechanical loops in instruments, stability under load and stability under thermal disturbance, will be considered for both static, or steady state, and dynamic, or transient, conditions.

8.3.1 Mechanical property groups

The mechanical functionality required of a loop structure in a precision system tends to involve the minimization of deflection under external or self-weight loads, the maximization of load capacity in small members, or, conversely, the maximization of elastic deflection in a flexure mechanism. Control of resonant frequencies may also be important.

Depending upon the strictness of geometric constraints placed upon the design, it may be best simply to select for load capacity on the basis of high Y, which could be taken as maximum allowable stress rather than always yield stress, and for low deflection on the basis of high E. As noted in Section 8.2, self-weight deflection of beams is minimized by selecting for high values of E/ρ and other similar ratios are useful in specific situations. Further, the resonant frequency of a beam of fixed geometry is proportional to $(E/\rho)^{1/2}$. In most situations, decoupling unwanted vibrations from a system is made easier if its resonances are at high frequencies, so for both static and quasi-static reasons E/ρ is chosen as a representative grouping.

Elastic mechanisms exploit linear strain and so a material of low E might seem preferable. However, it is normally a key feature of these mechanisms that the flexures behave extremely repeatably. Hence the deflecting sections must remain well within the linear elastic region, with no hint of local yielding being acceptable. The maximum strain is limited by the yield strength

$$Y = E\varepsilon \tag{8.7}$$

and, for this application, selection for high Y/E is desirable.

Resonance was considered above as a quasi-static constraint since its influence was as a noise source in a static system. However, it is treated as dynamic if it is being directly exploited, for example in a sensor system. Material selection would still depend on E/ρ, but whether a high or low value is best will depend upon the application. The other main dynamic effect, probably less commonly significant in precision devices, is inertial loading caused by accelerations. As a characteristic example, the stress, σ, due to centripetal acceleration in a uniform rod rotating at speed, ω is

$$\sigma = \frac{\rho\omega^2 L^2}{2} \qquad (8.8)$$

and so to obtain high rotational speeds (or other high accelerations) high values of Y/ρ should be chosen.

8.3.2 Thermal property groups

Thermal disturbance to a loop may be impressed either by environmental changes or by localized internal or external sources such as motors or the heat from an operator's hand. We can hope that environmental changes do not occur too rapidly, making it reasonable to assume that a body of moderate size will remain in thermal equilibrium with its local surroundings. In this case, the usual requirement is to place a limit on the direct thermal expansion. If a body is free, as in the case of a gauge block, or located kinematically, it may expand freely so selection for low α is best. However, often a body is more tightly constrained mechanically, perhaps forming one side of a rigid structure. In the limiting case, no change of dimension is possible. Forces are generated at the constraints that cause compression in the member exactly compensating for its attempted thermal expansion. If it remains in the elastic region, the stress generated by a temperature change, ΔT, under this condition is, for homogeneous, isotropic materials,

$$\sigma = E\alpha\Delta T \qquad (8.9)$$

Since as a general guideline stresses should be minimized, particularly in a metrology loop, selection for a low value of αE becomes important. Note, in passing, that internal stresses between phases are crucial to the low expansion behaviour of materials such as Invar or Zerodur.

With other heat sources, other properties should be considered. If a structure carries a dissipative device, perhaps an actuator, then it may be conducting an approximately steady heat flow. Taking the simplest case of Fourier's law for conduction along a uniform rod of length L and cross-sectional area A, carrying a heat flow, q, implies a temperature differential between the ends such that

$$q = \frac{kA\Delta T}{L} \qquad (8.10)$$

If the end distant from the source is at ambient, there will be a uniform temperature gradient towards the source and the average rise in the rod will be $\Delta T/2$. The expansion of the rod will be

$$\varepsilon = \frac{\Delta T}{2} \alpha = \frac{qL}{2kA} \alpha \qquad (8.11)$$

Under this condition it is preferable to use a material with a low value of α/k if the member is minimally constrained or, following the earlier argument, $\alpha E/k$ if it is heavily clamped.

Should there be a non-steady thermal disturbance, then normally a rapid return to equilibrium conditions is desirable. Again taking the simplest case of the one-dimensional heat equation

$$\frac{\partial \theta}{\partial t} = \frac{k}{C\rho} \cdot \frac{\partial^2 \theta}{\partial x^2} \qquad (8.12)$$

where θ is temperature relative to ambient and t and x are respectively time and spatial distance. The time behaviour is clearly an exponential decay from a disturbed state at a rate governed directly by the diffusivity $k/C\rho$. For rapid response a high value should be chosen.

The diffusivity controls the time taken for a disturbance to decay, but it has no influence on how the transient affects the system stability. As usual, it is expansion that is to be avoided. Non-uniform temperature profiles could be imposed on a loop in two extreme ways. One is to impress a defined profile through the application of a large heat source, the other is to inject a fixed quantity of heat that results in a local temperature rise. The first approximates a device being handled for a significant period of time. Although the temperature will be non-uniform, the expansion still depends on the average rise and so we select a material with low α or αE as appropriate. The second case is close to what happens if a motor operates for a brief period. A fixed amount of energy is injected into the structure and its temperature rises proportional to its volume specific heat. This rise results in expansion so now the material should best have a low value of $\alpha/C\rho$, again with E included for clamped members.

These are not the only mechanical effects that have temperature dependence, although they are probably the ones most often encountered. The variation of elastic constants with temperature will not be discussed here, but can be of some importance in sensor design, see the set of three articles by Gitlin, 1955.

8.3.3 Material profiles

The property groupings developed above are not, of course, the only ones for which a case can be made. They are, however, typical of those needed in instrument design. They are sufficient to demonstrate how arguments for others can be developed and they cover a good range of practical situations. The purpose here is to illustrate how the sifting of large amounts of information can be usefully systematized, not to provide a hard and fast algorithm that provides a 'correct' solution. Whether using a table of values or, ergonomically better, a graphical representation, humans can cope simultaneously with only a restricted number of parameters. No more than about ten should be used in any one selection exercise. In fact, the following discussion involves eleven groupings, and even then some compromises are needed to make choices from those already suggested.

A profile for a material is formed by noting the pattern of the end points of bar-graphs (or thermometer style displays) for each of the chosen property groups placed side by side. As before, logarithmic scales are used since orders of improvement are to be judged. The other chart systems did not normalize or non-dimensionalize the axes since there were no natural ways of so doing. Here several scales need to be compared and it is much easier to read them if they are all normalized in a standard way and then aligned. This is done by the simplest method available. The value of each grouping for any material is divided by the value of the same grouping for some standard material. The choice of standard is arbitrary; if a design group uses mainly steel they may normalize to steel so that better or worse than normal performance is seen in a way particularly meaningful to them. In general, there is some psychological danger in setting up any one material as the standard. Consequently a model material is proposed that has properties on the good side of typical in all headings, as shown in Table 8.2. The model has simple numeric values for its properties arranged so that the majority of materials will have normalized groupings in the region 0.25 to 4. Numbers outside this range attract attention as being 'unusual' and the graphical presentation should reflect this.

PROPERTY	VALUE	TYPICAL OF
Young's modulus	200 GPa	Steel
Maximum strength	300 MPa	Steel
Density	4000 kg m^{-3}	Ceramic
Thermal expansion	7 x 10^{-6} K^{-1}	Ceramic
Thermal conductivity	150 W m^{-1} K^{-1}	Brass
Specific heat	750 J kg^{-1} K^{-1}	Metals

Table 8.2: Properties of the model material for group normalization

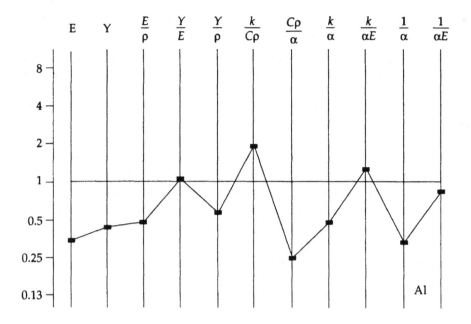

Figure 8.3   A property group profile typical of aluminium

Figure 8.3 illustrates the use of these rules to create a profile, using aluminium as an example. It must be emphasized that the figure shows a series of parallel single axis graphs. The line joining the values is there merely for visual emphasis of the pattern and has no physical or mathematical significance.

Groupings normally selected for low values are plotted as reciprocals so that higher points on the profile consistently indicate better performance. The order of presentation of the groupings must remain fixed if profiles are to be rapidly compared. The order should be chosen so that properties that are commonly similar for families of materials are close together. This tends to provide less jagged profiles which are easier to compare visually and adds emphasis to any materials that are different in just one respect. As constructed here, factors influencing mechanical performance are placed to the left and those relating to thermal performance are to the right. Also those more relevant to dynamic and transient behaviour are placed centrally with the steady state ones to either side.

The profile for aluminium gives an immediate impression of a material that is not very good, but rarely very bad. It is the sort of material chosen because it is easy to work with in applications that are not at the limits of performance. As expected, it shows well for diffusivity and, less obviously, for clamped members under thermal load. Otherwise, its high expansivity and modest stiffness and strength lead to mediocre ratings. A critical point must be raised before lending too much weight to the profile: what values should be taken for the basic properties. In the case of metals and metallic alloys, most properties are quite consistent but there is a little variation between different reference sources; most properties are somewhat temperature dependent. Other classes of materials show much wider spreads of values. All that can be said generally is that typical values should be used for screening and more specific profiles constructed for closer scrutiny if needed. Without doubt, the widest variation is in quoted strength, so this serves well as an illustration. Following earlier arguments, it is reasonable to expect that no yielding is tolerable in ductile materials used for precision applications. Yield strength is therefore an appropriate measure, although there is a case for taking some fraction of it to give practical margins. To stress how much care must be taken with strength parameters, consider again Figure 8.3. Adding just a few percent copper to the aluminium has little effect on most of the profile, but can increase the yield strength by a factor of three or more. The common alloy duralumin contains 4% copper and in many general workshops it is this that is used as 'aluminium'. In similar vein, the strength values given for both magnesium and titanium relate to typical, readily available alloys. Sometimes only an ultimate tensile strength is quoted and for ductile materials this over-estimates usable strength. A rough rule that estimates an effective yield at around 2/3 the tensile strength might be used to compensate. For brittle materials, which are unlikely to be used in direct tensile applications, compressive strength is usually good and the four point bending strength a good practical choice.

In many alloys, such as duralumin mentioned above, the extra strength is obtained at the cost of providing a material that is not generally in internal thermodynamic equilibrium. This must be true of any material that age-hardens but applies also more widely. There must be some questions regarding their dimensional stability and careful heat treatments may be needed to obtain good performance from them.

The usefulness of these selection aids can be judged from the following sections. They are most effective when used with a spreadsheet like computer database with a graphical interface so that rapid searching and redisplay may be exploited. Typical values for a range of materials that might be encountered in precision engineering are given in Tables 8.3 and 8.4, appended to this chapter. To emphasize that they are intended for guidance and not for use in design calculations without checking original data, only values for basic properties and groupings that have been normalized to the values in Table 8.2 are given, and then with truncated precision. Values have been collected from a variety of commonly available sources, sometimes with a degree of averaging between them. Exact sources are not cited since that would tend to give unwarranted authority to the numbers quoted. In some cases it is so hard to find consensus values that gaps have been left in the tables. Various handbooks, for example Bolton, 1989, show the detailed reasons for the existence of so many slightly different alloys and indicate less easily quantified parameters such as machinability, but they tend not to cover some properties and materials of interest for precision mechanical systems.

8.4 Materials properties

8.4.1 Metals

The tables include a range of pure metals and alloys, but show few surprises. The common machine-shop metals, steels, aluminium alloys and brasses, are much used as structural materials in precision engineering. With the exception of the strength of hard steels they are generally indifferent performers across the whole range of groupings. They are used because high performance is not always needed, they are readily available and have low costs for materials or manufacturability. It is mildly surprising that neither aluminium nor copper alloys give good overall thermal performance, largely because of their high expansivity. The low density of aluminium is sometimes useful. The copper alloys also have rather high densities. However they are easy to work and still find a major role when non-magnetic structures are wanted. Bronzes and, particularly, beryllium copper perform well in the category most associated with spring design, Y/E, with other properties at best moderate for most precision applications of springs. Their common use for this purpose perhaps reflects both the unacceptability of steels (magnetic fields and corrosion may reduce repeatability) and the ease of providing thin sheets for ligament systems, for which the requirements are rather different to those for energy-storing spring applications. With careful heat treatment, hard beryllium copper may have a very finely distributed precipitate that rapidly arrests the motion under stress of crystal dislocations so that it makes very repeatable flexure hinges. This illustrates the limitation of numerically based selection methods, for it is difficult to quantify and tabulate such a property. Steel and cast iron, which also offers usefully high damping, still dominate for larger structures such as machine tools. Further details on thermal properties are given by Gitlin, 1955a.

Amongst less commonly used metals, molybdenum appears as a good general purpose choice. It is closer to the model than other candidates, as seen from the profile given in Figure 8.4. Tungsten is also good overall except for its very high density. Neither is easy to work, but both should be considered seriously, especially for small devices where their inherent costs may not have too serious an effect on the total system cost. Molybdenum has been used, for

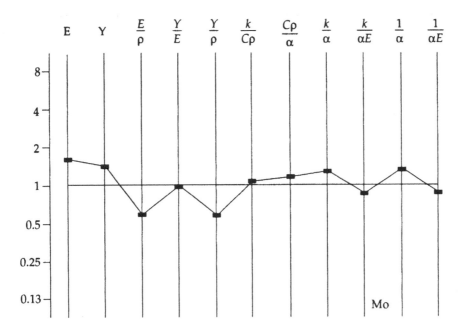

Figure 8.4 A property group profile typical of molybdenum

example, in the reference loops of scanning tunnelling microscopes, Fein *et al.*, 1987.

Beryllium and titanium are often thought of as high performance materials. The tables to some extent demonstrate why this should be. Beryllium is relatively stiff and has very low density but its other mechanical properties are not outstanding. It has a good combination of high specific heat and conductivity that make it attractive in some thermal environments. However, its reputation was not made in the field of precision engineering and, given its toxicity, it is best avoided unless E/ρ is dominantly critical. A similar verdict may be drawn for titanium. It has low density and, if not particularly stiff, shows good strength. As good properties are preserved at elevated temperatures, its attractiveness for some aerospace applications is clear. However, its profile shows that its only high point for instrument design is in quasi-static mechanical applications. Its thermal characteristics are unusual for a metal, but poor in our context. There are likely to be few cases in precision engineering that need its particular combination of good features enough to tolerate its weaknesses. In many respects its thermal performance is more like that of a ceramic.

Specialist alloys such as Invar are designed for specific purposes and perform well against the appropriate criteria, but tend to be at best moderate elsewhere. It is out-performed thermally by some other modern materials, but they are brittle ones. Amongst metals it dominates low expansion criteria and is strongly preferred for instruments needing high thermal stability in the presence of potential shock loads. Manufacturability is broadly similar to stainless steels, but it needs careful heat treatment to maintain its low expansivity since the phase mix must be just right.

8.4.2   Ceramics and glasses

Taken as a class, ceramics and glasses tend to be brittle, which restricts their application in systems carrying tensile and bending loads. However they have some very attractive properties and modern manufacturing techniques, which can control surface cracking, ensure that their use will increase. They can be formed by powder metallurgy processes or by low energy casting or moulding methods to produce complex shapes and, glasses particularly, can relatively easily be given very smooth surfaces.

The ceramics of most interest here tend to be oxides, carbides or nitrides, often with aluminium or silicon. They are generally quite strong, brittleness notwithstanding, rather stiffer than most metals and of reasonably low density. They tend to have poor thermal conductivity and moderate expansion

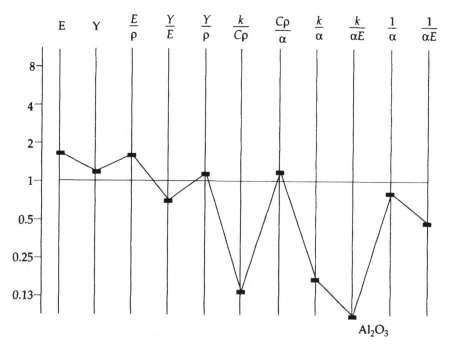

Figure 8.5   A property profile typical of a high density alumina ceramic

coefficients, although there are exceptions. Silicon carbide, which shows semiconducting properties at elevated temperatures, and tungsten carbide, which in practice usually involves a metal (cobalt) matrix, have higher conductivities. Alumina and zirconia are a little different and best suited to slightly different applications but show typical behaviour for a structural ceramic. Figure 8.5 illustrates this. They tend to be better than the model material on mechanical properties, but poorer, sometimes spectacularly so, on the thermal ones. Alumina type ceramics have been used successfully for major structural components such as small machine tool beds. The design rules are different, but the final result compares favourably with cast iron beds, Ueno, 1989. Their main application probably remains, for the present, small, light and strong elements to provide electrically insulating structures and for sintered parts with intricate profiles.

Silicon carbide, see Figure 8.6, and nitride are both high performance mechanical materials that maintain good to acceptable thermal characteristics, especially in the steady state. The relative difficulty of working them may well be worthwhile in challenging applications.

The relatively complex crystalline phases of many ceramics is rarely a problem in mechanical applications, but should not be totally ignored. For example, silicon carbide can cause optical phase changes at different parts of a surface if it is polished for use as an optical flat, an application for which it

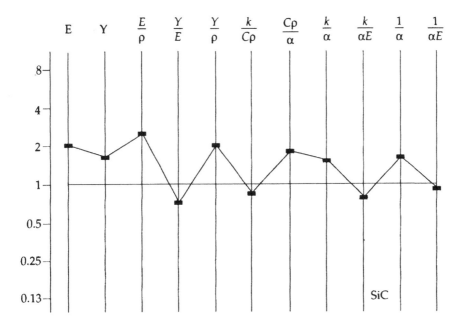

Figure 8.6 A property profile typical of silicon carbide

would seem very attractive mechanically. Also, whether produced by variations of slip casting or sintering, ceramics might be expected to have some porosity. However, densities of 99% or better of theoretical are available as standard and suppliers claim they can have zero porosity in the sense of gas leakage.

Conventional glasses have obvious applications in optics and optically worked surfaces can be very useful elsewhere, for example as mechanical datum surfaces. Generally their mechanical properties can be summarized as rather like a low strength aluminium, except for their brittleness. They are generally poor on the thermal criteria, although the expansion coefficients are quite small. Typical uses apart from optics are as electrically insulating substrates or as a matching material in a mixed loop. They can be moulded and machined reasonably easily. Fused silica and quartz are broadly similar to other glasses except that they have very small expansion coefficients, lower than is usually achieved with Invar. Very low αE values are obtained, which can be attractive in itself. However, their major feature does not show in the set of groupings. They have exceptional long-term dimensional stability. Quartz may also be extremely homogeneous. Such properties, coupled with the use of optically worked surfaces wrung together to produce almost perfect joints, leads to perhaps the best achievable metrology loop structures. This approach is only used when it is really needed, for example in the telescopes and rotors of the gyroscope relativity experiment, Everitt et al., 1979, since it is a difficult manufacture and assembly process.

Glass-ceramics are intermediate materials that generally look like and can be processed like glasses, but consist of a mixture of crystalline and amorphous phases. Controlling the phase structure allows them to be tuned to provide special properties. Of most interest here are the ultra-low expansion forms which after suitable annealing have expansion coefficients of no more than a few parts in 10^8 per degree over a small temperature range. The mechanism is again one of volumetric changes in different phases compensating for normal expansion. 'Zerodur' is the version that dominates telescope and other precision applications and the name tends to be used generically. It is well characterized for them, for example Brehm et al. 1985, although it has not been available for long enough for its temporal stability to be fully established. It is certainly good in this respect, but with such a complex structure it would be surprising if it were as good as silica. Large metrology frames on diamond turning machines and co-ordinate measuring machines now sometimes use it. The applications for such a material are indicated by the fact that it is nonsensical to attempt to compare it by graphical profiles. It is so extreme, in groupings involving α with some values over 100, that if thermal stability is critical it is well worth using even if so doing requires design complexities to allow for its poorer features.

8.4.3 Polymers and composites

Conventional polymers are little used in critical parts of precision machines because they have low stiffness and strength, poor thermal properties, including very large expansion coefficients, and poor dimensional stability. They tend to suffer from viscoelastic creep and also to absorb water from the atmosphere with a subsequent volume change. Elastomers are occasionally used to provide gentle spring retention and their hysteresis losses when load cycled have application in damping.

While bulk uses are limited, thin polymer layers are important system elements, particularly for bearings, see Chapter 9, and as adhesives. Polyacetal and polytetrafluorethylene bearing pads are much used on metal counterfaces for low and moderately loaded slideways.

Special designs with optically worked glass counterfaces give extremely precise translational behaviour, Smith, 1987. Both solid polymer and sintered polymer bronze composites can be used. Modern electroactive polymers have properties that may be of interest for this type of application and for providing thin film coatings with highly unusual property mixes.

Polypyrroles and other types can be electroactively polymerized onto substrates with electrical conductivities and surface morphologies that can be controlled over a wide range by electrical means. There is as yet little evidence of how they might be used, but the technology is growing rapidly and novel applications are almost certain to arise.

There is almost no good quality information about the behaviour of adhesive films in high stability structures. Undoubtedly they have relatively large dimensional changes from both temperature and humidity effects, but if they are used as thin films this may be tolerable. If, say, their linear expansion is 1000 times that of Zerodur, providing the film is only a few micrometres, the expansion of the glass will dominate if the dimensions are more than around 10 mm. On the other hand, this same observation suggests that differential behaviour in glued joints around a metrology loop might degrade its performance noticeably. The use of adhesives is also likely to cause some stressing as they cure.

There is anecdotal evidence that real problems exist with adhesive films in the most demanding instrument applications, and the optically wrung structures used by Everitt *et al.*, 1979, were used because of a lack of trust in adhesives under large environmental changes.

We have noted already the use of bimetals, which are composite materials, and further data is given by Gitlin, 1955b. Here we concentrate on non-metallic, mainly polymer ones. Common composites such as glass reinforced epoxy can provide a good lightweight covering shield with considerable strength and good acoustic damping, but are not otherwise much used in our context.

Carbon fibre reinforced plastics (CFRP) are more interesting since they offer potentially very high E/ρ values and good strength. Also, being composites, they can have very marked anisotropy, controllable by the alignment of fibres during manufacture. For example, the combination of mixed materials and strong anisotropy means that CFRP may have a negative thermal expansion coefficient in one direction. By careful adjustment, it is in principle possible to select for a zero expansivity in a single preferred direction. The applications to actuators and drives are clear for it offers combinations of stiffness, strength, density and toughness superior to silica and glass ceramics while having similar expansion properties in the operational direction.

Polymer concretes are another modern composite family of interest for large structures such as machine beds and worktables. They can contain the same aggregates, normally natural stone, as conventional concrete. Relatively conventional concrete mixes have been used as bulk structural materials with good damping for tables and in machine tools, Rahman and Mansur, 1990. However, usually the cement is replaced by a thermo-setting resin such as polyester, polymethylmethacrylate or epoxy. Natural granite could be considered as a composite and its use for tables and slides is based on its stability, ease of polishing and relatively favourable mechanical and thermal expansion properties.

Probably the most widely used commercial polymer concrete is 'Granitan', an epoxy and granite particle composite, that attempts to preserve the best properties of the rock while giving much easier and wider application at lower cost. It is another material for which the property profile method breaks down since it appears mediocre in its best categories. Again this is because it was designed for specific functions. Its E/ρ value is similar to metals that might be used for bulk structures and simple strength may not be a limiting factor in many designs. In any case, perhaps compressive strength, about five times higher than bending strength, should be used in the profile, given its design applications.

One major attraction is that it provides good damping and so produces machine frames with good vibrational characteristics. Its other advantage is that it can be cast, cold, into complex shapes with steel inserts for threads, guides and so on. Its expansion coefficient can be adjusted in the mix to match many metals. Polymer concretes may also be used as infill in welded metallic tubular structures, so increasing stiffness and damping without a vast increase in weight. In such applications, tensile strength of the concrete is almost irrelevant since a new level of composite with the metal carrying tension is produced, just as in conventional reinforced concrete. Although not the answer to every machine base design problem, they are a fine example of materials designed to facilitate economic, efficient precision engineering, Renker 1985, Weck and Hartel 1985.

8.4.4 Non-conventional materials for small devices

Materials such as glass ceramics and electropolymers might well be considered non-conventional and are certainly of interest to designers of small systems. This brief section is concerned more with materials that, for various reasons, can only be used in small devices or that can be used in unconventional ways at small scale.

Single crystal silicon is certainly an exceptional material and perhaps only just fits into this section. It is grown in large boules for the microelectronics industry as crystals that are almost defect free and of incredible purity. It is possible in principle to produce a complex artifact of a size of the order of a 100 mm cube that is one single crystal. As a single crystal, its properties are somewhat anisotropic and it is best to orient the lattice suitably for the function of a component. Its profile, Figure 8.7, which uses for Y the bending strength of the (111) close packed planes shows it to be a good material for mechanical and thermal applications.

Two profiles are shown in the figure, corresponding to data from two sources. One would expect less variation with a metal, but this nicely illustrates the precision with which the profiles should be used. Silicon is very brittle and appears to be perfectly elastic at room temperature. Thus it can provide

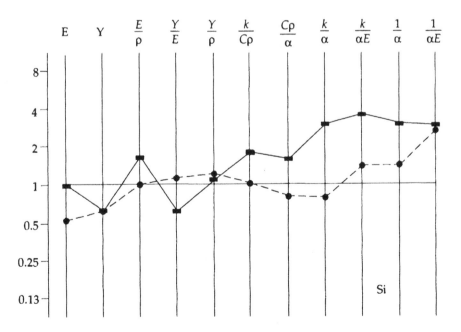

Figure 8.7 Typical property profiles for single crystal silicon

extremely repeatable and linear spring mechanisms that will definitely break and *not yield* if overloaded. Although it can prove expensive, this second feature can be very useful in standardization work. Also because it is a single crystal there is no inter-grain friction and so it has very low damping, making it useful for resonance sensors. There are well established polishing and etching technologies and it can be diamond ground with care. Usually a final etch is needed to remove surface cracks although its brittleness is not as big a problem for manufacture as is sometimes thought. Small structures can also be grown, for example by chemical vapour deposition, to build up miniature devices using extensions of the standard microelectronic fabrication technology. When growth techniques are used, the silicon is almost invariably polycrystaline (often called polysilicon to stress this). On the small scale, properties seem little different and a wide range of micro-mechanisms have been reported, see for example, Tang *et al.*, 1989 or Muller and Howe, 1990.

Tungsten is a metal not immediately or obviously associated with miniaturization for it is difficult to process consistently. However, it has the most useful property that it will grow directly and with a very good match onto silicon substrates. It is being increasingly used for very small devices made by a combination of growth and selective etching. Other metals are also grown to produce micro-components, typically by the LIGA process in which relatively thick polymethylmethacrylate coatings are selectively etched and, for example, nickel grown into the recesses, Guckel *et al.*, 1991.

Thin layer deposition, even sometimes to atomic monolayers, is now relatively routine in many materials. Apart from its simple applications, unusual composites can be constructed by layering normally incompatible materials on top of each other. The processing is very expensive, but it yields potentially unique property mixes. Examples are left to the imagination of the reader, for we are entering a new world of materials design.

Finally there arises the use of diamond, possibly the ultimate material, for small structures. Its hardness is legendary and its other properties superb for much precision engineering. Only cost and the difficulty of working it (and neither is in the slightest way trivial) prevent its use. Considerable strides have been made in the last few years in growing diamond-like and apparently pure diamond films onto substrates such as silicon carbide and molybdenum. Carbon rich gasses in a reducing atmosphere can be induced to deposit carbon and, because the surface energies of the graphite and diamond allotropes are quite different, the process energy can be controlled to allow diamond to deposit while any graphite formed is etched away again, Lifshitz *et al.*, 1989, Komanduri *et al.*, 1990. It is now becoming conceivable that either a composite including a diamond layer or a whole structure of diamond, a substrate having been etched away, could be quite feasible design options for extreme performance small devices.

8.5   Materials for transducers

We include here for completeness brief notes acknowledging some of the influences of modern materials on transducer design. This is a rich field since, almost by definition a transducer requires materials with relatively high sensitivity to some disturbance, be it only high expansivity for making thermometers.

Many special materials have been designed for transducer use, several of which are examined in Chapter 7. Mechanical examples include piezoelectric ceramics and polymers, iron-rare earth alloys such as Terfenol that have very high magnetostrictive coefficients, and shape memory metals which switch between two grossly different shapes over a temperature change of a few degrees. In all three cases their induced change of dimension can be used as an actuator to provide controlled displacement. When used in precision devices, they effectively become structural elements attached to loops and so are subject to all the mechanical and thermal loads discussed. It is unfortunately common to discover that their overall property profiles are rather poor. To exploit their unusual behaviour we must design around the overall weaknesses, perhaps by supporting them by higher performance structural materials to form a relatively complex actuator subsystem.

Single crystal silicon has generally good but not exceptional properties. It is unusual for a different reason. It can be exploited in high precision flexure designs because of its virtually perfect elastic-brittle behaviour. It is a semiconductor for which there is a very advanced microelectronics technology. Thus smart sensors in which mechanical elements and sensing and conditioning electronics are incorporated in a single monolithic structure are quite feasible. Miniature diaphragm pressure sensors and simple accelerometers for the automobile industry are the most common examples but more complex devices are also being produced for a wide variety of applications. The processing technologies developed for silicon electronics devices can be exploited to make small, or even moderate sized, mechanical devices. Various alkaline and organic etchants are strongly anisotropic, attacking perhaps hundreds of times faster along the close packed crystal planes than across them. This allows the production of deep narrow channels and high selective directed features. Other etchants are very sensitive to the presence of dopants. The combination of the growth of new layers, diffusion of dopants and various etchings enables the production of complex three dimensional structures, including deep, controlled undercuts, Petersen, 1982.

Beryllium has mechanical properties that might be exploited. It is stiff and has very low density, so the speed of sound in it is unusually high and perhaps of use in acoustic or vibration sensors. Its Poisson's ratio is even more extreme, being below 0.1. Very stiff actuators have been constructed by exploiting the expansion of a rod due to the Poisson effect when a hydrostatic collar around it is pressurized. Using parallel rods of beryllium and a more normal metal within one pressure system might provide a super stiff differential actuator with almost common path action.

8.6 Case studies

Choice of material can never be isolated from other design decisions, yet for present purposes it is convenient to act almost as if it were possible. We shall, therefore, not look at real design problems but at the outline of archetypal ones.

8.6.1 Simple flexure mechanism

Flexure mechanisms are often preferred for providing precise translational motions over distances of up to a millimetre or so. Consider the application of a simple mechanism consisting of a small moving platform mounted to a larger static base through two parallel flexing legs. The device is characterized by dimensions of around 20 mm. To avoid drift at joints, and to save assembly costs, a monolithic construction is wanted.

A set of functional requirements can be established. It must have good repeatability, implying a material with good dimensional stability and low hysteretic losses. As much range as is realistically possible is wanted, so the material must be able to sustain considerable stress to allow a large degree of flexure. It is judged that a high resonance will be helpful in reducing susceptibility to vibration, so low mass and high mechanism stiffness are wanted. Thermal stability is preferred but not critical as there are no internal heat sources and the environment will be reasonably controlled. Note in this context that the device is effectively free in a direction parallel to the legs and more strongly constrained in a direction perpendicular to them.

General mechanical design can do much to satisfy the needs for stiffness and strength, providing that space is not too highly restricted. Referring back to Chapter 4 or Equations 8.1 to 8.4, increasing the depth, d, of the flexures rapidly increases overall stiffness with little weight penalty. However, the deflection that can be obtained without exceeding a specified maximum stress decreases linearly with the depth. The flexure breadth linearly affects both stiffness and weight without influencing maximum deflection. Almost always, some combinations of these can be found to provide suitable overall parameters without overly restricting material choice. It will usually be possible to choose a material less than ideal in stiffness or strength if other factors favour it.

From the earlier discussions, deflection relates to Y/E and resonance to E/ρ. To increase stiffness, it would be useful to have fairly high E and selecting high

Y/E implies also high Y. Because of this and to emphasize that low mass would be helpful, it might be appropriate to place a weaker preference on a high Y/ρ. Thermal demands are not high but perhaps some attention should be paid to both $1/\alpha$ and $1/\alpha E$ since some environmental influences are likely. An outline profile to act as a template for screening candidate materials can now be sketched, Figure 8.8. The numeric values against which marks are placed are less important than the relative positions, other than that they should be high enough to show that selection for high values is wanted.

Glancing down a table or set of profiles, approximate matches on the chosen columns are readily found. Diamond, as always, scores well but can be discounted. Several ceramics, silicon carbide and silicon nitride (particularly), show well on the mechanical side and are satisfactory to good on the secondary thermal constraints. Even tungsten carbide looks interesting, although mainly because of its extremely high strength. They will not be rejected at this stage, but their brittleness and relative difficulty of manufacture are noted. Silicon also comes into this category. Of metals, tungsten does well except on E/ρ and titanium looks good mechanically and moderate for the thermal characteristics. Hard steels are modest thermally but quite good otherwise. Mild steels, aluminium and copper alloys all are modest overall with beryllium copper and duralumin best on the critical Y/E criterion. Perhaps slightly surprisingly,

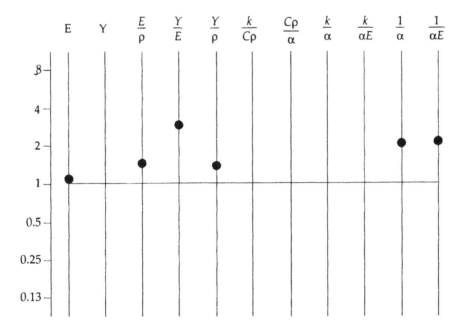

Figure 8.8 Profile template for a small flexure mechanism

beryllium copper, which is commonly used for springs, is not particularly good in any other category for this application.

Overall, single crystal silicon is almost certainly overkill, but silicon nitride and carbide could be serious contenders for high performance devices. As noted earlier, brittleness might be an advantage in such systems. For less demanding applications steel and aluminium alloys are probably to be preferred because of their ease of use. The higher density, poorer overall thermal characteristics and cost probably outweigh the small advantage that copper alloys show relative to duralumin. On a small piece, for easiest machining, duralumin remains a perfectly sensible choice even though its profile does not at first look all that appealing. Clearly, it is important that a first search through materials options is not ambitiously narrow.

8.6.2 Talystep revisited

The interaction between the availability of materials and instrument design strategy in Talystep and its successors was used as an illustration in Section 8.1. How do the decisions leading to the development of the 'Nanostep', Garratt and Bottomley 1990, appear in the light of the selection guides discussed here?

The essential problem in the design is to provide a metrology loop, with dimensions of around 300 mm, with a small measurement gap that is stable to considerably better than 1 nm over periods of a few minutes and, preferably, does not drift much over longer periods. Stiffness is clearly one critical feature of the loop. As there is no specific constraint on the outer geometry of the loop, almost any desired value can be achieved by choosing suitable dimensions for the members and there is no materials selection constraint implied. Even a very low E would probably be acceptable if there were compensating advantages elsewhere. The loop is quite large so self weight may have a minor influence on the selection process. Since the instrument will remain in one place, gravitational distortion is not a particular problem unless it leads to unacceptable creep. Neither E/ρ nor Y/ρ are critical in this respect. High resonance is probably useful and, since the loop is quite large, a high E/ρ has some advantage. A material with high internal damping would be good, but not at the cost of reduced dimensional stability. For the short term stability, thermal effects will almost certainly be dominant. Extremely low loop expansions are the ultimate design requirement. Since the instrument will be touched quite often, a high diffusivity to give fast settling is desirable. Otherwise, the only important internal heat source is an intermittently used motor which can be placed relatively far from the critical members. Environmental effects are probably the more significant thermal disturbance. It should be possible to design the loop largely on kinematic principles so that most expansions are nearly free, in which case the critical selection is for high $1/\alpha$., with a weaker preference for high k/α. The longer term critical issue is temporal stability. The key points in the profile template are shown in Figure 8.9.

Zerodur type glass ceramics have very low diffusivity but are reasonably good on the mechanical constraints, with E/ρ better than for many metals. However, the expansion related groupings, which are critical here, are so favourable that all else fades to insignificance. The fact that a non-uniform temperature gradient takes a long while to decay does not matter at all if it causes no noticeable expansion. The C_p/α rating is extremely high, so the structure will cope will thermal load if necessary. By comparison, aluminium is a poor choice. It must be stressed that this explains the changes that have occurred between the designs. The materials available for loops of this size and for the precision bearings in the traverse were very different in the 1960s, when the choice of aluminium was easily defended. Looking at other options in case they offer alternative design approaches, Invar is quite good except for diffusivity. It certainly does not compete with glass ceramic but might have been considered for Talystep. The costs would have been much higher and not justified for the advantage gained. The other metals generally good for loops, such as molybdenum or tungsten, would never be seriously considered for this application. Silicon carbide might well be considered seriously now were it not for Zerodur. Ceramics of this type come out well overall and can be used successfully for medium sized structures even though they have low damping.

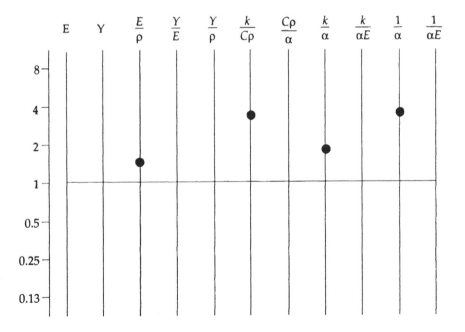

Figure 8.9 Profile template for an extreme stability metrology loop

	E.Mod	Strength	Density	Sp.Heat	Th.Con.	Expans.
	E	Y	ρ	C	k	α
Aluminium	0.35	0.40	0.70	1.2	1.6	3.4
Berylium	1.6	1.2	0.50	2.4	1.3	1.7
Copper	0.65	0.80	2.2	0.50	2.6	2.4
Magnesium	0.25	0.50	0.45	1.4	1.0	3.7
Molybdenum	1.6	1.5	2.6	0.35	0.90	0.70
Titanium	0.60	2.3	1.1	0.70	0.15	1.3
Tungsten	2.1	4.5	4.8	0.20	1.1	0.65
Cast iron	0.75	0.70	1.8	0.70	0.35	1.6
Mild Steel	1.1	1.0	2.0	0.55	0.35	1.6
Steel, spring	1.0	2.0	2.0	0.55	0.35	1.6
Steel, hard	1.1	3.3	2.0	0.55	0.25	1.6
Invar	0.70	1.3	2.0	0.65	0.15	0.15
18/8 S.S.	1.0	1.7	2.0	0.70	0.10	2.3
Elinvar	0.85	1.2	2.0	0.60	0.07	0.55
Brass 70/30	0.55	1.5	2.1	0.50	0.75	2.8
Bronze 90/10	0.65	2.0	2.2	0.50	0.35	2.4
Phos. Bronze	0.55	1.7	2.2	0.50	0.45	2.4
Be. Copper	0.65	2.5	2.1	0.45	0.65	2.4
Duralumin	0.35	1.0	0.70	1.2	1.0	3.3
Silicon(111)	0.95	0.65	0.55	0.95	1.1	0.35
Diamond	6.0	10.0	0.90	0.50	3.9	0.15
Si. Carbide	2.1	1.5	0.80	1.3	0.85	0.55
Si. Nitride	1.6	3.3	0.80	0.75	0.20	0.50
Alumina	1.7	1.2	0.95	1.4	0.25	1.2
Zirconia	1.0	2.7	1.4	0.65	0.02	1.5
Tung. Carb.	3.6	11.0	3.8		0.35	1.0
Fused Silica	0.35	0.25	0.55	1.1	0.01	0.07
Fused Quartz	0.35	0.25	0.55	1.1	0.01	0.07
Crown Glass	0.35	0.25	0.65	0.95	0.01	1.1
Zerodur	0.45	0.30	0.65	1.1	0.01	0.01
Granitan	0.15	0.07	0.65	1.3	0.01	1.7

Table 8.3: Basic properties of some materials of interest in instrument design, expressed relative to the model material of Table 8.2.

NORMALIZED PROPERTY GROUPS

	E	Y	E/ρ	Y/E	Y/ρ	$k/C\rho$	$C\rho/\alpha$	k/α	$k/\alpha E$	$1/\alpha$	$1/\alpha E$
Aluminium	0.35	0.40	0.50	1.1	0.60	1.9	0.25	0.45	1.3	0.30	0.80
Berylium	1.6	1.2	3.4	0.70	2.5	1.2	0.65	0.80	0.50	0.60	0.35
Copper	0.65	0.75	0.30	1.2	0.35	2.3	0.50	1.1	1.7	0.40	0.65
Magnesium	0.25	0.50	0.50	2.2	1.2	1.7	0.15	0.25	1.2	0.25	1.2
Molybdenum	1.6	1.5	0.65	0.95	0.60	1.1	1.2	1.3	0.80	1.4	0.85
Titanium	0.60	2.3	0.55	3.9	2.1	0.20	0.60	0.10	0.20	0.80	1.3
Tungsten	2.1	4.5	0.45	2.2	0.95	1.3	1.3	1.7	0.85	1.6	0.75
Cast iron	0.75	0.70	0.40	0.95	0.40	0.25	0.80	0.20	0.30	0.65	0.85
Mild Steel	1.1	1.0	0.55	0.95	0.50	0.35	0.70	0.25	0.20	0.65	0.60
Steel, spring	1.0	2.0	0.50	2.0	1.0	0.35	0.65	0.20	0.20	0.60	0.60
Steel, hard	1.1	3.3	0.55	3.2	1.7	0.20	0.70	0.15	0.15	0.65	0.60
Invar	0.70	1.3	0.35	1.8	0.65	0.10	8.5	0.90	1.2	6.4	8.8
18/8 S.S.	1.0	1.7	0.50	1.6	0.85	0.07	0.60	0.04	0.04	0.45	0.45
Elinvar	0.85	1.2	0.40	1.5	0.60	0.05	2.2	0.10	0.15	1.8	2.1
Brass 70/30	0.55	1.5	0.25	2.9	0.70	0.70	0.40	0.25	0.50	0.35	0.70
Bronze 90/10	0.65	2.0	0.30	3.1	0.90	0.30	0.45	0.15	0.20	0.40	0.65
Phos. Bronze	0.55	1.7	0.25	3.0	0.75	0.45	0.45	0.20	0.35	0.40	0.75
Be. Copper	0.65	2.5	0.30	4.0	1.2	0.70	0.40	0.25	0.45	0.40	0.65

Table 8.4: Property group values for the materials in Table 8.3, expressed relative to the model material of Table 8.2. (cont. over...)

NORMALIZED PROPERTY GROUPS

	E	Y	E/ρ	Y/E	Y/ρ	k/Cp	Cρ/α	k/α	k/αE	1/α	1/αE
Duralumin	0.35	1.0	0.50	2.7	1.4	1.2	0.25	0.30	0.80	0.30	0.85
Silicon(111)	0.95	0.65	1.7	0.65	1.1	1.9	1.6	3.1	3.3	3.0	3.2
Diamond	6.0	10.0	6.9	1.7	11.5	8.8	2.6	23.0	3.8	5.8	0.95
Si. Carbide	2.1	1.5	2.7	0.75	1.9	0.80	1.9	1.6	0.75	1.8	0.90
Si. Nitride	1.6	3.3	1.9	2.2	4.2	0.40	1.2	0.45	0.30	2.0	1.3
Alumina	1.7	1.2	1.7	0.70	1.2	0.15	1.1	0.20	0.10	0.85	0.50
Zirconia	1.0	2.7	0.70	2.7	1.9	0.02	0.60	0.01	0.01	0.70	0.65
Tung. Carb.	3.6	11.0	0.95	3.1	3.0			0.35	0.10	0.95	0.25
Fused Silica	0.35	0.25	0.65	0.65	0.45	0.02	8.5	0.20	0.55	14.0	40.0
Fused Quartz	0.35	0.25	0.65	0.65	0.40	0.02	8.6	0.15	0.40	14.0	40.0
Crown Glass	0.35	0.25	0.55	0.65	0.35	0.01	0.50	0.01	0.02	0.90	2.5
Zerodur	0.45	0.30	0.70	0.70	0.50	0.02	97.0	1.5	3.3	140.0	308.0
Granitan	0.15	0.07	0.30	0.40	0.10	0.01	0.45	0.01	0.04	0.60	3.3

Table 8.4: (continued) Property group values for the materials in Table 8.3, expressed relative to the model material of Table 8.2.

References

Ashby M.F., 1989, On the engineering properties of materials, *Acta metall.*, **37**, 1273-1293.

Ashby M.F., 1991, On Material and Shape, *Acta metall.*, **39**, 1025-1039

Bolton W., 1989, *Newnes Engineering Materials Pocket Book*, Heinemann Newnes. Oxford.

Braddick H.J.J., 1966, *The Physics of Experimental Method*, 2nd ed., London:Chapman and Hall.

Brehm R., Driessen J.C., v. Grootel P. and Gijsbers T.G., 1985, Low thermal expansion materials for high precision measurement equipment, *Precision Engineering*, **7**, 157-160.

Charles J.A. and Crane F.A.A., 1989, *Selection and Use of Engineering Materials*, London:Butterworths.

Chetwynd D.G., 1987, Selection of structural materials for precision devices, *Precision Engineering*, **9**, 3-6.

Chetwynd D.G., 1989, Materials selection for fine mechanics, *Precision Engineering*, **11**, 203-209.

Everitt C.F.W., Lipa J.A. and Siddall G.J., 1979, Precision engineering and Einstein: the relativity gyroscope experiment, *Precision Engineering*, **1**, 5-11.

Fein A.P., Kirtley J.R. and Feenstra R.M., 1987, Scanning tunneling microscope for low temperature, high magnetic field and spatially resolved spectroscopy, *Rev. Sci. Instrum.*, **58**(10), 1806-1810.

Garratt J.D. and Bottomley S.C., 1990, Technology transfer in the development of a nanotopographic instrument, *Nanotechnology*, **1**, 38-43.

Gitlin R., 1955a, How temperature affects instrument accuracy, *Control Engineering*, April 1955, 70-78.

Gitlin R., 1955b, Compensating instruments for temperature changes, *Control Engineering*, May 1955, 70-77.

Gitlin R., 1955c, How temperature compensation can be used, *Control Engineering*, June 1955, 71-75.

Guckel H., Skrobis K.J., Christenson T.J., Klein J., Han S., Choi B., Lovell E.G. and Chapman T.W., 1991, Fabrication and testing of the planar magnetic micromotor, *J. Micromech. Microeng.*, **1**, 135-138.

Komanduri R., Fehrenbacher L.L., Hansen L.M., Morrish A., Snail K.A., Thorpe T., Butler J.E. and Rath B.A., 1990, Polycrystaline diamond films and single crystal diamonds grown by combustion synthesis, *Ann. C.I.R.P.*, **39**(1), 585-588.

Lifshitz Y., Kasi S.R. and Rabalais J.W., 1989, Subplantation model for film growth from hyperthermal species: Application to diamond, *Phys. Rev. Letts.*, **62**(11), 1290-1293.

Lindsey K., Smith S.T. and Robbie C.J., 1988, Sub-nanometre surface texture and profile measurement with NANOSURF 2, *Ann. C.I.R.P.*, **37**, 519-522.

Muller R.S. and Howe R.T., 1990, Technologies for microdynamic devices, *Nanotechnology*, **1**, 8-12.

Petersen K.E., 1982, Silicon as a mechanical material, *Proc. I.E.E.E.*, **70**(5), 420-456.

Rahman M. and Mansur M.A., 1990, Development and evaluation of ferrocement legs for a lathe, *Int. J. Mach. Tools. Manufact.*, **30**, 629-636.

Renker H. J., 1985, Stone-based strutural materials, *Pre. ision Engineering*, **7**, 161-164.

Smith S T, 1986, PhD Thesis, University of Warwick.

Tang W.C., Nguyen T.H. and Howe R.T., 1989, Laterally driven polysilicon resonant microstructures, *Sensors and Actuators*, **20**, 25-32.

Ueno S., 1989, Development of an ultra-precision machine tool using a ceramic bed, *5ᵗʰ Int. Precision Engineering Seminar*, Monterey, CA.

Waterman N.A. and Ashby M.F., 1991, *Elsevier Materials Selector*, Elsevier, London

Weck M. and Hartel R., 1985, Design, manufacture and testing of precision machines with essential polymer concrete components, *Precision Engineering*, **7**, 165-170.

9 SLIDEWAYS FOR LONG RANGE PRECISION MOTION

Since rolling element bearings always inject some load fluctuations and noise, they can rarely be used in ultraprecision translation systems. Thus we nearly always apply a variation of a slideway involving the motion of a set of bearing pads sliding over a datum surface. This chapter explores their kinematics and so proposes some general rules for both geometric and materials design. It then looks in more detail at different pad principles. Relatively little, other than an overview, is given on fluid film bearings, for which extensive literature exists elsewhere. A rather fuller treatment of solid contact systems is given. In all cases, practical limits of precision are discussed.

Introduction

In Chapter 7, we discussed a range of phenomena of direct use as actuators with open loop capability at the nanometre level. Almost invariably their range was limited to a few tens of micrometres. Guide mechanisms rarely range over more than a few millimetres for the best elastic designs. To achieve a linear traverse over many millimetres or a continuous rotary motion some form of slideway mechanism must be used. For a considerable while some machine tools have had slides of sub-micrometre repeatability and surface metrology instruments ones repeating to around 10 nm over tens of millimetres of traverse. Recent advances in the tribology of bearing way systems have resulted in slideway motions having a repeatability of better than 1 nm over a 50 mm range. The bearing systems used in these applications are usually of one of three types:

1. Air or oil hydrostatic

2. Hydrodynamic or elastohydrodynamic.

3. Dry polymeric or lamellar solid lubrication

A fourth category, rolling element linear bearings, generally constructed as pre-loaded crossed-roller tracks, is rarely used in such applications although they are much used in optical and some other instruments. Their particular combination of load capacity and inherent noise places them outside the scope of this chapter. Each of the types listed above has its own merits and disadvantages which will influence the magnitude and sources of error expected from it. Before discussing specific types of bearing it is useful to examine some error sources common to almost all of them.

9.1 Sources of error in bearing slideways

We shall explore generalized error sources in terms of finding some ideal rules for design.

Sliding bearings consist generally of a stationary 'way' upon which is mounted the moving carriage. This concept is illustrated in Figure 9.1 which shows a standard vee groove and flat slideway geometry commonly found in machine tools. It illustrates a number of common error sources.

First, and most obviously, the drive axis is not coincident with either the centre of friction or the centre of inertia. Driving through an axis that is not coincident with the centre of inertia will result in a torque perpendicular to the axis that is proportional to the mass and the acceleration of the carriage and to the square of its offset. When a steady state velocity has been achieved, this effect will not be significant and it is often ignored with continuously driven machinery. However, systems using stepper motors or involving discontinuous motion may be performance limited by this error. The second

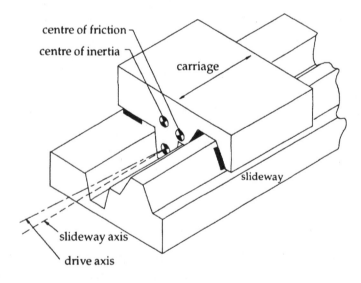

Figure 9.1   A typical slideway geometry

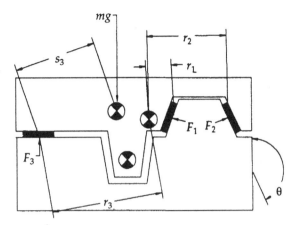

Figure 9.2   Cross-section of the slideway shown in figure 9.1

condition will induce varying moments perpendicular to the drive axis proportional to the frictional drag and the offset. Because of the fickle nature of the friction phenomena, the magnitude is likely to be unpredictable, but the effect will be present even in the steady state. Hence:

Rule 1. Try to ensure that the centres of inertia and friction are coincident with the drive axis. Note that this rule is a corollary of the Abbe alignment principle. In this form it has also been called the Bryan alignment principle.

The position of the centre of inertia can be determined with high accuracy using any one of several numerical or experimental techniques, for example by finding the point of balance. Finding the centre of friction for a particular slideway configuration is not as simple. It lies at the point where the vector sum of the products (moments) of the frictional force at each bearing pad and its distance from that point equates to zero.

As an example, consider the cross-section of the slideway of Figure 9.1 shown in Figure 9.2. If it is assumed that the load force is concentrated in the middle of the bearing pads, then the moments about both the centres of friction and inertia, M_f and M_I and the total resolved forces are given by

$$M_I = s_1 \wedge F_1 + s_2 \wedge F_2 + s_3 \wedge F_3 = 0 \tag{9.1}$$

$$M_f = r_1 \wedge \mu |F_1| \hat{k} + r_2 \wedge \mu |F_2| \hat{k} + r_3 \wedge \mu |F_3| \hat{k} \tag{9.2}$$

$$F = mg = -(F_1 + F_2 + F_3) \tag{9.3}$$

where μ is the coefficient of friction at each pad and the vectors r and s correspond to the positions of the centroids of the bearing pads.

The right hand side of Equation 9.2 contains three unknown forces, F_i, and vectors, r_i. Although the slide is correctly constrained for a single freedom semi-kinematic mechanism, there are too many unknowns to enable the centre of friction to be found. Some assumptions must be made. In fact, already some have been, for the same coefficient of friction has been allocated to each pad. If, further, we assume that, for example, the two reaction forces to the right hand side of the centroid are equal in magnitude then we can find all the reaction forces. Once these are known, the equations can be solved for the centre of friction.

As well as relying on a large number of assumptions, such procedures still leave the problem of arranging for the centroid to lie on the desired axis. For higher precision designs, alternative constructions can provide variation of the force applied to individual bearing pads. Not only does this allow some tuning of the slideway characteristic, but calculation of the centroid is considerably simplified. Figure 9.3 shows an example of such a compensated design. Plain (or boundary) lubricated bearings traverse the slideway and define the motion while the normal force is applied by oil hydrostatic bearings on the opposite faces. As a consequence, the relative size of the frictional forces can be controlled. See Section 9.3 for further discussion.

Figure 9.1 also shows the axis of the drive not parallel to the axis of the carriage motion. This causes a tendency for the carriage to twist which must be prevented by extra forces at some of the bearing pads. So:

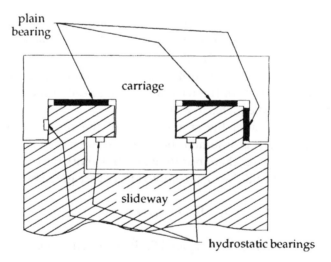

Figure 9.3   A high stiffness bearing using opposed forces

Rule 2. Try to ensure that the drive and slideway axes are colinear.

Although this may have only secondary effects on the straightness of a properly designed slideway, the increased and variable loads on bearing pads increases internal stresses and may induce unnecessary wear. Variation of the forces may also shift the position of the centre of friction. They should be avoided as a matter of principle in good design. Whenever possible, alignment datums for the drive system are machined at the same time as the slideway itself so that both are subject to the same machine setting errors. Some slideways will be made from materials that are not readily machinable and must, in consequence, be aligned with the drive after manufacture. Techniques for coupling the drive to the carriage with minimum off-axis influence have been discussed in Chapter 5. Another way of attenuating non-rectilinear forces is to use the way system itself. For example, the slideway of Lindsey *et al.*, 1989, used a slave carriage to drive the specimen carriage via non-influencing couplings (Chapter 5). A repeatability of better than 1 nm in the transverse plane was achieved over displacements of up to 50 mm. This system will be discussed further in Section 9.5.

Rule 3. Optimize the tribological properties for the purpose of precision.

The main requirement in most slideway designs is for a reliable and robust system. However, for ultra-precision designs, repeatability and linearity are the parameters of key importance. The former is limited by the thermal stability and tribology of the components and bearings. As noted in Chapter 8, materials selection is a compromise between bulk cost, manufacturability, thermal and mechanical properties. More detailed tribological considerations will be discussed later in this chapter. The friction between the carriage and way system results in forces which cause distortions of its structure and so affect linearity. If these forces are constant, then so too will be the induced distortion. Such error sources can be compensated after characterization by, for example, the method of reversed measurement, Section 3.7.2. Wear of the slide will cause a change in the way shape and thus invalidate any compensation. As a consequence the carriage bearing is usually constructed from a soft material or non-contact methods, such as hydrostatic bearings, used. In many routine applications, soft bearings are used sacrificially to ease maintenance: here the search for high performance outweighs such concerns. Soft bearing materials may pick up and embed any hard particles suspended in the atmosphere or surrounding fluids. The counterface may then be abraded and the original datum lost. This can only be avoided by encapsulation or careful filtering of all environments within the system.

In designs that involve material contact, friction forces will vary with displacement due to stick-slip and non-uniform abrasion or delamination of the bearing surface. Softer interface materials are likely to have higher friction coefficients and wear rates (techniques for varying these characteristics are discussed in Section 9.4). Stick-slip induces random errors into the motion and changes in the susceptibility of a bearing to it are also unpredictable. Because the variations in frictional force are usually proportional to the normal force, a low friction coefficient, as found with hydrostatic bearings and some polymeric bearings under high load, is desirable. At high load, even low coefficients may cause significant force variation. Thus in most precision applications some increase in friction coefficient will be tolerated in order to use a material with little tendency to stick-slip behaviour.

Rule 4. Minimize the energy dissipated at or passed through the bearing interface.

Work done in maintaining carriage motion will create an equivalent amount of heat at the bearings that must eventually be extracted again. In many cases this simply involves allowing the heat to diffuse into a large thermal mass or into the environment. For many engineering applications thermally induced errors are not highly significant. This is not the case in precision systems, where thermal loading, particularly if varying dynamically, can be a major error source. Apart from the direct frictional effects, thermal loads are commonly introduced by motor drives, machine coolants and the flow of fluid in hydrostatic bearings. The problem of heat from motor drives can be dealt with by thermally and mechanically decoupling them to reduce conductive, convective and radiative transfer. The heating effect of a fluid may be reduced by regulating its temperature at the input stage to be that of the room in which the instrument or machine is housed. In practice, close temperature regulation of fluids is far from simple and control to better than 0.1 K can be very expensive. Another alternative that is becoming cheaper, is to use low expansion ceramics such as 'Zerodur' for the structure. This approach assumes that thermal expansion causes the motion errors and often this is so. It will not, however, help if, for example, thermal effects cause softening of a polymeric bearing, since there will be no good conduction paths to take the heat away.

Rule 5. Make the transverse stiffness of the slideways bearing as high as possible to reduce susceptibility to vibration and other perturbing forces.

The purpose of any slide is to convey an object from one place to another. In doing so, the force distribution within the system will almost invariably alter. This happens as the carriage moves under the line of action of a machine tool or contacting probe and as the centre of gravity of the carriage itself moves. The importance of ultra-stiff designs for systems of the highest precision is emphasized by noting that a conventionally 'stiff' bearing of, say, 20 MN m^{-1} can be re-expressed as having a compliance of 50 nm N^{-1}. The area of spherical, and many other types, of mechanical contact varies with load according to a power law usually in the range 0.67 to 1, Archard, 1957. Consequently, the rigidity of many slideways may be increased by applying a preload. For example, in Figure 9.3. the preload can be directly varied by controlling the hydrostatic pressure of the bearings on the counterface.

Another common strategy is to use feedback to maintain the position of the specimen relative to a fixed datum. It could be applied to all six degrees of freedom of carriage compliance, but in practice, for reasons of cost, it is rarely used except where out of plane forces in no more than two of the possible freedoms dominate the error behaviour. The authors do not know of any cases in which it has been applied for complete geometrical control.

Stiffer slideway designs have reduced vibration sensitivity because the amplitude of motion is proportional to the square root of the ratio of the input energy to the stiffness at a given frequency. This becomes of primary importance in mechanisms having little internal damping. Related effects are discussed more fully in Chapter 10.

Rule 6. Try to observe the rules of kinematic constraint.

Chapter 3 outlined the general principles of kinematic constraint by considering the contact of points on infinitely rigid surfaces and noted that under low load conditions, spheres and small flats provide a reasonable approximation to the kinematic condition. For designs in which high preloads are applied the conditions for pure kinematic design clearly break down. However, the general principles need not be discarded entirely but merely modified slightly. In such semi-kinematic designs, bearings are allowed to distort elastically to provide larger contact regions. We then classify the centroids of those regions as the points of kinematic constraint. If the bearing ways are run-in to give a high conformity, the amount of distortion required to maintain a reasonable area contact is reduced to acceptable limits. Consequently, as more precision is used in component manufacture, it is possible to achieve very stiff component interfaces.

Rule 7. Optimize the kinematic design to minimize errors.

Kinematic analysis indicates that a single degree of freedom translation can be modelled as five contact points suitably placed on the ways. Each of these will closely follow the slide geometry, including its errors. The motion error of the carriage will depend on how the paths of the individual pads interact and we should attempt to place the pads so that the overall error is small. Consider the simple model in Figure 9.4, a much exaggerated schematic representation of two of contacts. We assume that a carriage moves a workpiece under a probe positioned *s* from the end of the way. The ideal way is a straight line parallel to the *x*-axis. The feet on the carriage are spaced *d* apart and the working surface

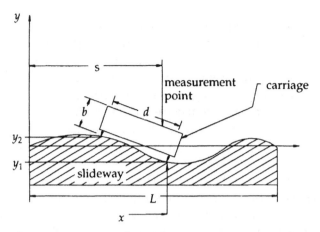

Figure 9.4   Exaggerated view of straightness error in way and motion

is b above the pads. Thus ideally the probe measures relative to a constant height b above the slide.

In reality, the way surface is not flat but has height variations about the x-direction given by $f(x)$, where x is the position of the leading foot. The effective height at the probe is then found by

$$y = y_1 - (x - s)\tan(\theta) + \frac{b}{\cos(\theta)} \tag{9.4}$$

where

$$\theta = \sin^{-1}\left[\frac{y_1 - y_2}{d}\right] \tag{9.5}$$

Here $y_1 = f(x)$ but $y_2 = f(x - d\cos(\theta))$ and so even if $f(x)$ were to be expressed in analytic form, a closed solution for the carriage position cannot generally be obtained. However, if the way errors are small compared to the pad separation, as is virtually always the case, we may approximate to find

$$y \approx f(x) - (x - s)\left[\frac{f(x) - f(x - d)}{d}\right] + b\left(1 + \frac{1}{2}\left[\frac{f(x) - f(x - d)}{d}\right]^2\right) \tag{9.6}$$

Various special cases can be discovered, but in general all three terms in Equation 9.6 will be detrimental. The first is just the way error, but the others involve tilts of the carriage and so are related to Abbe errors. The example illustrates why in general we should keep b as small as possible, control the magnitude of $(x-s)$ to stay as close to zero as function allows and space the bearing pads as far apart as is practicable. Note, however, that increasing d for a given traverse requirement implies increasing the total length of the slideway and so may contribute to an increase in its error.

9.2 Fluid film bearings

There are two main classes of fluid film bearings, that is bearings in which there is no solid to solid contact, at least during normal running. In both, a pressurized working fluid is introduced between the counterfaces. In a hydrostatic system the pressure is externally generated whereas a hydrodynamic bearing is internally pressurized, the relative motion of the bearing components supplying the pump action. A great variety of working fluids may be used, but only two are of significance in our context: air and low viscosity mineral oil. All types can provide good combinations of precision, low drag and transverse stiffness when well designed. Since, generally, higher fluid pressures can be obtained by external pumping, the stiffest designs tend to be hydrostatic. Conversely, the absence of external pumps tends to make hydrodynamic systems less expensive for any given level of precision.

High quality bearing design has many subtleties that are not appropriately addressed here. We shall content ourselves with a rather over-simplified outline of their principles that provides just enough detail to allow an appreciation of their capability. Beyond this we make use of examples of their application. There are highly specialized texts on many aspects of their design, but for an overview in somewhat greater depth than we provide general texts on tribology, such as Arnell *et al.*, 1991, or design, such as Slocum, 1992, are probably more useful.

9.2.1 Air or oil hydrostatic bearings

The basic idea of a hydrostatic bearing is illustrated by Figure 9.5. Pressurized fluid is introduced into the gap between the pad and the way and the load is carried by the resulting film. At first sight it might seem that the load capacity is the product of the pressure and the pad area, with an overall bearing stiffness to match. However the use of an internal reservoir alters this behaviour pattern. Fluid is supplied to the pad through a restrictor from a high pressure source so that an approximation to constant flow is obtained (using a true constant delivery pumping system would be better, but much more expensive). Under static conditions the reservoir pressure will be such as to support the load and to drive the requisite flow through the gap between the pad lands and the counterface (usually this is the bearing way). The clearance will adjust to this balance. If the load increases, the gap tends to close and so to maintain the flow

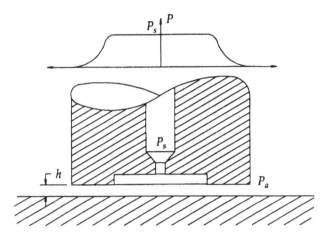

Figure 9.5 A schematic view of a hydrostatic bearing pad and its operational pressure distribution

the internal pressure rises rapidly. The extra load is thus supported with almost no net change in the gap, that is a very high transverse stiffness is obtained, even with bearings having conveniently large clearances.

The actual behaviour of hydrostatic bearings involves a complex interaction between several design parameters and the overall stiffness between carriage and slideway is related to the interfacial pressure by some power law. The stiffness per unit area of bearing, λ, depends upon the fluid supply pressure, P_s (for low flow rates), the shape of the delivery orifices and the clearance between the carriage and slideway, h, in a relationship of the form (see for example Elwell and Sternlight, 1960, or Laub, 1960)

$$\lambda \; \propto \; \frac{P_s}{\eta h^m} \tag{9.7}$$

where η is the viscosity of the fluid and m is a constant, usually about 4, or slightly higher, dependent upon the design parameters.

For high stiffness, it is desirable to have both a very small film gap and as high a pressure as possible without inducing instabilities. Another approach is to clamp the carriage between two bearings which results in a doubling of the stiffness. This extends naturally to the idea of using hydrostatics to centralize a carriage. A cylindrical datum shaft will probably have a smaller error in the straightness of its axis than in its surface shape. It is also likely to be circularly symmetrical. So placing around it a collar with equispaced circumferential hydrostatic pads, which will self align to its axis, will probably provide a more precise motion. Additional pads can be placed along the axis to provide resistance against axial tilts. The use of distributed bearings is often referred to as compensation.

From Equation 9.7 it is apparent that for a given supply pressure and clearance, the stiffness is inversely proportional to the viscosity. Thus very rigid air (gas) bearings can also be produced. A recently constructed example uses two spheres mounted into closely fitting cups and attached to each other by a silica shaft. This design relaxed the requirement for axial colinearity of the bearings at each end of the shaft, but is deliberately overconstrained kinematically. The open loop stiffness of this bearing was approximately 0.16 N nm^{-1} axial and 0.07 N nm^{-1} radially (see Precision Engineering, April 1989, page 110). The extremely low viscosity of air results in correspondingly low viscous drag forces making it possible to drive the shaft with a low stiffness, and thus low-influencing, drive system.

The pad illustrated in Figure 9.5 has a recess for the distribution of high pressure fluid and thus higher load bearing capacity. This is usually at the expense of stability of the bearing which may in turn be increased by the use of a restrictor at the supply orifice as shown. The optimization of these parameters for a given design is beyond the scope of this book and the reader is referred to more advanced texts such as Gross, 1962, or Barwell, 1954. Attempts have been made to provide increased stiffness in hydrostatic bearings by local control of the pressure within the individual pads. Placing sprung flaps within the pad modifies its behaviour dynamically by a method loosely comparable to that used in servohydraulic systems.

By definition, hydrostatic bearings require a pressurized supply. This causes a continuous energy input additional to that dissipated by viscous drag. With oil based bearings, the temperature of the inlet can be regulated and the outflow used to extract heat so that the bearing and its associated structures remain close to thermal equilibrium. Being a compressible medium, the situation with air bearings is less simple in both principle and practice. There will always be adiabatic expansion of the gas as it passes through restrictors and when it exits the bearing. Local cooling of regions around the bearing ensues and is not readily compensated by adjusting the inlet temperature. Its effect can be reduced by maintaining a small gap, which may not always be desirable on other grounds. An additional problem within air bearings is Helmholtz resonance. This is caused by an elastic wave travelling between the exit and supply orifices in a manner analogous to the sounding of a wind instrument. Although it is generally acknowledged to exist, there is very little information regarding its minimization, which seems a well preserved proprietary secret of the manufacturers.

It has been proposed that gas be supplied through a porous journal instead of through a capillary or orifice, to provide higher intrinsic damping. Although porous type hydrostatic bearings have been in existence for more than fifty years, Sneck, 1968, there are few examples of use in ultra precision designs at the time of writing. A spherical segment rotary bearing with very high claimed performance has been reported which incorporates a precisely formed graphite concave element*. This provides a porous medium for the introduction of the gas (usually highly purified nitrogen) and also a low friction interface should touch-down occur.

In general, air hydrostatic bearings exhibit similar performance characteristics to their oil counterparts but the are more factors that can introduce noise and instabilities. They may be avoided by careful design but invariably at increased cost. Air-based systems tend to be lower cost because the flow paths are simpler. It will usually be easier to filter suspended particles from air than from oils so that there is less chance of them interfering with the bearing operation. Also air may simply be vented to the atmosphere whereas oil must be collected and recirculated. Although the compressibility of air may be a problem both in terms of adiabatic cooling and internal resonances, it may give lower coupling of vibration from the pumping system compared with an almost incompressible oil. Table 9.1 summarizes the major comments made above.

ADVANTAGES	DISADVANTAGES
Very low friction	High cost due to the precise machining tolerances and fluid supply
Low wear giving a long lifetime and excellent long term repeatability	It has an intrinsic energy source that can lead to instability
High stiffness per unit area	Design is complex and best left to experts
It is possible to use closed loop control to maintain positional accuracy	Pumps act as vibration sources

Table 9.1 A summary of the advantages and disadvantages of hydrostatic slideway bearings

---

* Produced by Cranfield Unit for Precision Engineering (CUPE), Cranfield, Bedfordshire, UK.

9.2.2 Hydrodynamic rotary bearings

The behaviour patterns now recognized as belonging to hydrodynamic, or self-pressurized, bearings was first described in the famous account of Beauchamp Tower and explained in 1886 by Osborne Reynolds. One of its easiest forms to visualize is a couette flow in which a flat plate moves through fluid above and at a small angle to a flat base. If the open angle faces forwards, fluid will be scooped in and, to preserve continuity, must accelerate to pass through the smaller exit aperture at the rear of the plate. A local increase of pressure must build up to provide this acceleration and incidentally it provides also a lift force on the plate. Intuitively, we see that the magnitude of the lift force depends on the angle of attack, the speed of motion, the viscosity of the fluid and the size of the plate. Since reasonable speeds must be maintained to ensure that the lift force does not drop to a value where metal to metal contact might occur, this phenomenon can only be used for precision bearings that are in continuous motion. Practically, this means that hydrodynamic bearings are restricted to unidirectional, continuous rotary applications of either journal or thrust bearings.

Even a perfectly round shaft running in an, inevitably, slightly larger round hole will generate an internal pressure in the presence of fluid. The transverse load on the shaft will cause it to be offset in the hole so that the radial difference creates a fluid wedge similar in action to the flat plate case described above. At very low speeds there is essentially continuous contact between the surfaces and consequently fairly high friction and wear. As speed builds up there is more tendency for oil to remain on the contacting surfaces and so friction drops. Eventually a regime is reached when there is almost complete lift-off and the bearing surfaces are kept apart by a complex interaction of elastic distortion of surface asperities and rapidly changing properties of the highly squeezed oil film. This is called elastohydrodynamic lubrication (ehl) and offers the lowest friction. It is difficult to design and then control real systems to remain consistently in this region. Further increases in speed cause complete separation of the bearing surfaces and a true hydrodynamic regime is reached. The friction is now essentially viscous drag and so will rise approximately linearly with speed. At sufficiently high speeds the fluid film will become turbulent and dissipate more energy so the friction will then rise more steeply with speed. The general behaviour for a given bearing design can be summarized by a Stribeck diagram, which plots friction against a non-dimensional group that reflects design pressures and rotary speed. The generally expected form is indicated in the modified Stribeck diagram of Figure 9.6.

The purely round bearing produces relatively little lift and cannot, of course, produce any thrust reaction at all. Much ingenuity has been given to improving

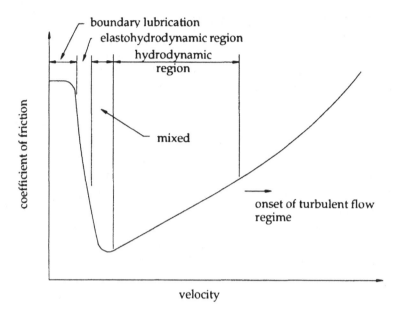

Figure 9.6   Friction of a hydrodynamic bearing of given dimensions and fluid at different speeds. (Modified Stribeck Diagram).

this state. For thrust bearings, one of the simplest methods is to make one of the bearing surfaces not flat, but stepped as a series of low angle wedges. Each then generates a couette flow lift force. Simple steps can also help and such designs seem originally due to Lord Rayleigh. A more subtle use of the same idea is to make one surface with spring supported pivoted plates. These will adjust their tilt relative to the moving face as the internal pressure changes and, with care, can provide good operation over a considerable range of driving conditions. All these ideas can also be applied to journal bearings, although making accurately profiled shapes on the circular, particularly the recessed, surfaces is less convenient than on the flats. Only the tilting plate designs are readily fabricated. An alternative approach is to provide a pumping action for unidirectional bearings by inscribing or etching spiral grooves into the bearing surfaces. Considerable pressures can be raised this way, but again obtaining good performance needs much care. Details of the design of these types of bearing are beyond the scope of this work. Readers are referred to works of increasing specialization and depth such as Cameron, 1976, Fuller, 1984, or Muijderman, 1966.

While both oil, air and other working fluids find considerable use in the broader engineering environment, there are relatively few applications in true precision machines. One special and highly effective use is in the provision of reference motions for roundness measuring machines such as the Rank Taylor Hobson Talyrond. The requirement here is for extremely well defined and precise circularity of running of a shaft that carries a micro-displacement sensor. The sensor detects the radial variation between its path and a stationary workpiece. The variation is assumed to relate directly to the roundness errors of the workpiece. The spindle rotates at constant speed, usually within a small factor either side of 10 rpm, and is lightly and predictably loaded since it carries only the sensor. The situation is ideal for a precision hydrodynamic bearing. All early designs, and some modern ones, use a vertical spindle supported by a hemispherical bottom bearing, 50 to 75 mm diameter, machined to high surface quality and precision of shape of around 25 nm. This is flooded in a low viscosity and very well filtered mineral oil. The upper end of the spindle is supported by an elastically mounted rolling element bearing in a kinematic arrangement. All the datum precision comes from the bottom bearing. Some bearings have had small sheet metal flaps attached to the rotor to simulate a wedge profile, but there is little evidence that this improves the accuracy compared to that of the simple shape. Similarly the absence of systems with spiral grooves suggests that they also are an unnecessary complication for this application. From the very first machines built in this style, the roundness error of the spindles has been in the 20 to 30 nm region. Modern systems tend to run to a slightly oval profile with similar levels of run-out and repeatability of the slightly non-circular path within a few nanometres, Chetwynd and Siddall, 1976. The error is probably due mainly to bearing components, but there may be some component caused by the rotation of the oil film relative to the bearing. Larger, high speed systems can suffer a failure mode known as half-speed whirl, which is caused by a pumping action of the fluid driving an eccentricity of the shaft concentrically around the journal at half of the rotary speed. At the highest precisions, perhaps a slight wobble due to related fluid effects may be present.

9.3   Lubrication with lamellar solids

A lamellar solid has a structure that is strongly bonded in one plane with relatively weak bonds between such planes. In suitable orientation they can carry direct stress while being quite weak in shear and so have potential as solid lubricants. The lamellar solids most commonly used to reduce the frictional coefficient between two surfaces are graphite and molybdenum disulphide. A very effective and simple way of applying these lubricants is to rub the counterface surfaces with a cloth impregnated with fine powder. The powder tends to stick to the surface, providing a layer of low friction material. The mechanism of adhesion is probably a combination of physical embedding by the hard edges of the thin plates and chemical bonding. To form a low shear strength lubricating layer, both materials require a relatively soft and reactive counterface. Consequently, only their application to metal surfaces will be considered. As there are contrasting mechanisms of friction when using graphite or MoS_2, each will be discussed separately.

9.3.1　Frictional behaviour of graphite

The anomalously low friction coefficient of surfaces after they have been rubbed with a graphite based powder was first investigated and explained from a crystallographic viewpoint by W. L. Bragg in 1928. A more comprehensive study was carried out and presented in a highly readable paper by Savage, 1948, which is still held to be a consistent explanation. It showed that the friction coefficient of graphite powders is strongly related to the presence of water vapour. Under light loads at atmospheric pressure, the coefficient of friction between graphite and graphite, or graphite and metal is low (approximately 0.1-0.2). If, however, the interface is outgassed in high vacuum, both the friction and wear rate increase dramatically. This is explained by the action of water molecules on the edges of planar graphite particles (see also Bowden and Tabor, 1964). The graphite edges are very highly reactive and water molecules are attracted by and bonded to them, forming an effective shield. When rubbed onto a metal surface the shield is disrupted by the high stresses and the exposed edges attract to the counterface surface. Because of the extreme hardness of graphite in the basal plane, the particles are able to both chemically bond and physically 'key in' to the surface. Further rubbing and absorption of water molecules from the atmosphere will result in an average alignment of the graphite platelets to the friction angle. Reversal of the motion will result in reorientation to align with the new angle.

The low coefficient of friction and wear rate is therefore thought to be a result of the low cohesive (shear) strength between individual plates in the presence of water vapour from the atmosphere and their very high hardness due to the close carbon-carbon bond.

The model has some implications for the lubrication of nanometre precision slideways. An unknown degree of damage may be caused during each reversal of slideway motion when the platelets reorient to the friction angle. Also it is unsuitable for high vacuum applications because outgassing will strip the shielding molecules from the graphite leading to high wear and friction. Finally, the stability of such a reactive material over long periods of time is uncertain. However, neither is there firm experimental evidence that graphite lubrication will not perform adequately at the nanometre level under low load conditions in a normal laboratory atmosphere.

9.3.2 Frictional behaviour of molybdenum disulphide

The low friction coefficient, commonly in the region 0.1-0.2, between two metal surfaces after rubbing with a fine molybdenite powder is again explained by the weak van der Waals attraction between cleavage faces and an extremely high hardness perpendicular to this direction, Deacon and Goodman, 1958, Pauling, 1945. Although the exposed edges of the plates are very reactive, they can form a stable, well adhered molybdenum oxide layer making these laminates inert to other contaminants: in a pure state it can even be hydrophobic. In consequence, it continues to exhibit good frictional properties in high vacuum and at temperatures of up to 350 °C, above which decomposition occurs.

The stability and vacuum compatibility of this material make it favoured as a thin film lubricant for nanotechnological applications. Although fears about the abrasive effects of the hard particulates during changes in direction of motion still exist, an MoS_2 bearing way for mapping and manipulation at the molecular level has been proposed, Evans et al. 1989, and will be discussed in one of the case studies at the end of this chapter.

9.4 Dry or boundary lubrication

Boundary lubrication is considered to occur when either the bearing load is too high or sliding velocity too low to enable the separation of surfaces by hydrodynamic effects. Under these conditions, the two counterfaces will contact at the highest surface asperities which will subsequently be compressed until there is sufficient contact area to support the load, Archard, 1957, Johnson, 1987.

Gross plastic distortion will result in a bonding and interlocking of the solids. This must then be overcome if one surface is to slide relative to the other. Clearly, under these conditions, the lowest frictional coefficient will correspond to the lowest interface shear strength. There are two distinct approaches to the reduction of the shear strength between two contacting solids. One is to coat the surfaces with a molecularly thin layer of liquid and the other is to isolate the two materials with a soft solid. These correspond to conventional boundary lubrication and polymeric bearings respectively.

While a repeatibility of the order of ten nanometres is achievable using hydrostatic and hydrodynamic bearings, it is generally accepted that, because of the atomic scale of the friction phenomena itself, dry solid or boundary lubricated bearings are likely to be better at the nanometre level. Most ultra-high precision slideway designs have used either boundary lubrication with metallic ways or polymeric pads with ceramic and other materials.

9.4.1 Liquid film boundary lubrication

Typical boundary lubricants consist of a thick viscous liquid containing a fatty acid. The liquid serves only to supply the fatty acid to the surfaces where the acid itself provides the lubricating effect. The lubricant typically consists of a long hydrocarbon chain with a polar head at one end. These heads readily bond to a metal surface to create a carpet of hydrocarbon molecules on it. Stearic acid ($C_{17}H_{35}COOH$) is one such molecule, the long chain having a carboxyl group attached at one end, Freeman, 1962. It is postulated that during sliding there is a combination of intimate metal-metal contact in small regions when high asperities coincide and thin film separation where the load conditions are less severe. Thus if the attached molecules are longer than the average asperity height the friction coefficient is mainly attributable to their shear strength. There still exists wear at the occasional contact points.

Boundary lubrication has a number of drawbacks for use in nanotechnological applications. First and foremost it can usually be applied only when one of the sliding counterfaces is metallic and quite reactive. (The surface reactivity of the metals can be enhanced by adding elements such as chlorine, sulphur or phosphorous to the lubricant). Unfortunately metals may have undesirable properties such as poor thermal stability, see Chapter 8. Also, the relatively high hardness of metals of higher dimensional stability will cause wear of both surfaces and consequent loss of a stable datum reference. However, there are many applications where the reference datum is relatively unimportant and this approach then provides an excellent high stiffness slideway with a coefficient of friction in the region 0.05-0.2. Another case study in this chapter discusses Miyashita's ultra-precision surface grinder which combines hydrostatic bearings for a high load and boundary lubrication (or hydrodynamic lubrication at higher velocities) for a high stiffness.

9.4.2 Polymer bearings

Since metals are not the most stable materials upon which to base the reference or datum of a high precision mechanism, newer ceramics and glasses such as Zerodur, Mexim, silicon and titanium diboride (which is nearly as stiff as diamond, but relatively cheap to produce) are increasingly being used. For ultra-precision mechanisms it is almost essential to use the low thermal expansion ceramics now available. The first requirement is that the slideway is smooth and straight, which may generally be achieved using standard optical polishing techniques. The slideway surface must then be traversed without introducing significant damage. This can be done by separating the two surfaces by a thin polymeric layer. Three materials are of interest for this type of application, ultra-high molecular weight polyethylene (UHMWPE), polyacetal and poly-tetra-fluoro-ethylene (PTFE). A major problem associated with this type of bearing is the provision of a film sufficiently thin to retain the stability of the measurement loop. Thickness must be limited because of the very high thermal expansion coefficient (approximately 10^{-4} K^{-1}) and creep associated with all of these polymers.

The evidence for the metrological fidelity of UHMWPE is mainly inferred from research into the wear properties of artificial hip joints in which a strong dependence of wear rate with surface finish has been observed. Dowson *et al.*, 1987, investigated a three millimetre pin rubbing against a stainless steel counterface having a surface finish value in the region 0.01 μm R_a. At a load of 100 N and a surface velocity of 250 mm s^{-1}, a typical wear rate was 10^{-6} mm^3 N^{-1} m^{-1}. This implies that a one metre traverse would cause a perpendicular displacement of 14 nm between distant points in the slideway and carriage. During these experiments the bearings were immersed in distilled water, probably to simulate the presence of sinovial fluids. This is likely to keep the bulk UHMWPE at a relatively low temperature (it becomes unstable at around 100 °C). The test conditions are extreme for the majority of applications and much lower wear rates would normally be expected, making nanometre level repeatability possible in principle. The friction coefficient is commonly comparable to boundary lubricated metals, in the region 0.1-0.2.

Providing a thin bearing film of UHMWPE is very difficult. Two feasible methods are; direct plasma polymerization, or to shave the polymer from bulk and then glue it to a substrate. Neither is straightforward. Preparatory procedures for gluing polyethylene are complex and often involve some form of chemical etching that will probably alter the tribological properties of the material, Lees, 1989. Plasma polymerization requires good process control with specialized and expensive vacuum deposition equipment, Boenig, 1982.

Polyacetal (or acetal for short) is a lot easier to obtain in thin film condition. A polymer-metal matrix is commercially available, for example in Glacier 'DX' bearings. A cross-section through a typical hemispherical bearing pad is shown in Figure 9.7. The polymer is held in place by a porous sintered bronze layer bonded onto a steel backing plate. The layer of polymer between the bronze and counterface can be reduced to approximately 20 μm by machining the excess acetal from the bearing face. Consequently, the bearing pad will consist of two thin metal layers, each approximately 0.3 mm, plus the polymer layer, a total thickness of less than 0.7 mm. The combined expansion coefficient is then approximately 11 nm K^{-1}, giving nanometre level stability in a well thermally controlled environment.

The coefficient of friction for acetal against a smooth glass or steel counterface is around 0.3 when dry, but drops rapidly to 0.1 or less in the presence of a suitable lubricant. Rubenstein, 1961, postulates that boundary lubrication occurs as a consequence of molecular adsorbtion. Hence, the

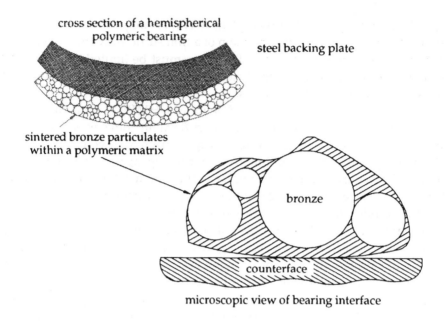

Figure 9.7 Cross section of a typical polymer bearing with the polymer keyed into a porous substrate

lubrication becomes more effective as the chain length of the lubricant molecules increases until an optimum is reached. Beyond this, if the lubricants are derived from a homologous series, the high viscosity tends to significantly increase the drag force. Adsorbtion and absorption of the lubricant may also cause plasticization of the polymer surface. This will have two effects. Firstly, the shear strength of the surface will reduce to give a lower friction coefficient, but also an increase in wear by film transfer mechanisms (for a more complete discussion of wear mechanisms in polymers see Lancaster, 1984, Atkins, 1984, Senior and West, 1971). Secondly, the additional compliance will enhance hydrodynamic effects due to the localized distortion of the surface (that is, the elastohydrodynamic effect), Tabor, 1954. Detailed experiments on acetal indicated that the elastohydrodynamic effect may be significant at velocities of 0.1 mm s^{-1} or less, Fort, 1962. Many stylus and other probe instruments operate with scans of this order of speed, and acetal bearings may cause excessive variability of lift. It was further shown that high creep rates associated with most polymers may contribute to an increase in the friction coefficient at lower speeds sometimes to the original dry friction values. Interesting results have been obtained in 'dwell' tests on this material. A dwell test consists of running the polymer bearing with a minute amount of fluid (0.5 µl or less) applied to the interface. The friction coefficient is then monitored as a function of the traverse distance to ascertain the endurance of the lubricant film. It has been found that application of either a 20 cst silicone oil or di-2-ethylhexylsebacate lubricant will reduce the wear rate by two orders of magnitude to approximately 10^{-6} mm^3 N^{-1} m^{-1} and can persist for many kilometres of travel without a discernible increase in the friction coefficient from a value of approximately 0.1, Lancaster, 1984.

PTFE is of particular interest for precision slideways. It consists of straight, long chain, fluorocarbon molecules having a high stiffness in all directions and low intermolecular adhesion perpendicular to their axes. This produces an almost perfect combination of properties; a high stiffness normal to the plane of sliding and a low shear strength parallel to it. However, the bulk material has undesirable features such as very low strength, a high creep rate and a high thermal expansion coefficient. Lubrication is achieved by the gradual deposition and spreading of the film similar in nature to the spreading of a deck of cards. A number of characteristics significantly affect its performance as a bearing material. At high loads or velocities, the drag forces may be sufficient to tear crystals or grains from the matrix resulting in a high wear rate, Mackinson and Tabor, 1964. This increase in wear rate has been observed by Tanaka, 1973. It is almost universally agreed that the sliding of PTFE will induce at the surface a thin oriented film likely to be no more than 400 nm thick, Steijn, 1968, Tabor, 1982. Electron diffraction studies have shown it to be highly oriented, with the axis of the molecules parallel to the traverse direction. Observations of how the film bridges surface scratches show a tendency for it to split parallel to the direction of traverse, a phenomenon not unexpected in view of the weak intermolecular adhesion in this direction, O'Leary and Geil, 1967, Steijn, 1968, Tanaka *et al.*, 1973. Surface tension effects then draw the stretched film into fibrils.

The low thermal diffusivity of PTFE means that it is likely to retain the heat generated by friction. Unless this can be conducted away, large temperature changes will occur. The resulting interface stresses are accentuated by the large thermal expansion anisotropy and a dilatation due to a phase transformation that occurs at around 19-25 °C. Both result in gross delamination of the film. The expansion of the film is far greater perpendicular to the molecular orientation and so correspondingly aligned bubbles may grow. They will eventually buckle and tear off forming wear debris, Hornbogen and Karsch, 1983, Lhymn, 1986. Wear rates have been reduced by exposing the polymer to γ-radiation. This also increases the coefficient of friction and hardness of the material, McLaren and Tabor, 1965, Briscoe *et al.*, 1987.

The low stiffness and thermal diffusivity can be artificially increased using the metal composite bearing configuration similar to that for acetal, Figure 9.7.

If PTFE is the matrix, then the overall thermal diffusivity will be roughly the sum of the polymer film value and that of bronze. A very thin film will be able to conduct significant heat to the bronze with only a small temperature drop across it. This heat will then be conducted away by the high diffusivity of the bronze substrate. The stiffness is almost independent of the extremely thin layer and can be considered to be nearly that of the bronze. As the load increases to higher values the thin film will cease to enlarge its real contact area with the counterface and the pressure in the polymer will necessarily increase. Because the shear strength of the polymer varies slowly with pressure, Tabor, 1982, the friction coefficient of PTFE appears to *reduce* with increasing load. Typically, a well designed bearing will give a friction coefficient between 0.2-0.08. The addition of a small amount of lead also tends to enhance the characteristics of this type of bearing. It is not clear how this works but it may encourage adhesion of the polymer film and increase the thermal diffusivity of the matrix.

It is speculated that wear of the film is compensated by the drawing of a fresh layer of polymer from the matrix. Examination under an optical microscope reveals the bronze particles as a series of isolated dots surrounded by the polymeric matrix. Comparison of these with clean bronze shows that they have been slightly darkened by the polymer film. This can be verified by abrading the pads with a file and viewing the bronze that has been exposed. If the bearing is then rubbed on a smooth surface and viewed again, the darkening becomes apparent, further confirming that the film thickness is only a small fraction of a micrometre.

9.5 Case studies

There are currently research projects throughout the world on ultra-precise machine tools for diamond turning and ductile regime grinding applications. Instruments have been produced for over a hundred years that necessarily operate with nanometre precision, see for example Rowland, 1902. The following case studies have been chosen specifically to illustrate these themes. Two concern large scale translation of specimens in instrumentation systems requiring nanometre or sub-nanometre resolution. They indicate the variety of solutions that may be applied to nominally similar design criteria. The third examines the slides of a grinding machine of extreme static stiffness.

9.5.1 Nanosurf II

The Nanosurf II stylus instrument was developed at the U.K. National Physical Laboratory for the measurement of surface finish and profiles of x-ray

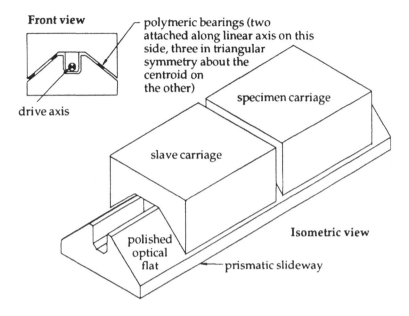

Front view

polymeric bearings (two attached along linear axis on this side, three in triangular symmetry about the centroid on the other)

specimen carriage

drive axis

slave carriage

Isometric view

polished optical flat

prismatic slideway

Figure 9.8   Schematic of the way system in Nanosurf II

microscope mirrors, Lindsey *et al.*, 1988. A prismatic slideway provides the traverse surface for two carriages that rest on it, held only by gravitational force. About this is a frame to which a stylus-based displacement transducer is attached so that specimen surfaces can be measured as they traverse below it. To reduce the thermal expansion of the measurement loop, all of these components are constructed in the low expansion ceramic Zerodur. Ignoring the stylus transducer and drive couplings, a schematic diagram of the slideway system is shown in Figure 9.8. One carriage serves as a slave to transmit forces from a micrometer drive to the specimen carriage along the axis of the prism (in accordance with design rule 2, Section 9.1). This operates by using a non-influencing drive coupling at each end of the slave carriage to attenuate any out-of-axis forces from the drive before they are transmitted to the specimen carriage. More precisely manufactured drives and couplings would alleviate the need for this, but only at the expense of higher manufacturing cost.

Attached to the underside of each carriage are five PTFE bearing pads, two to one side and three to the other. Assuming that the coefficient of friction for each pad is the same, then the centre of friction can be readily calculated (this is not strictly true since the friction coefficient will probably vary with load). However, there is only a 2:3 difference in loading between pads on each side, so the change in friction coefficient is unlikely to be large. The carriage bearing forces can be modelled from the force diagram of Figure 9.9.

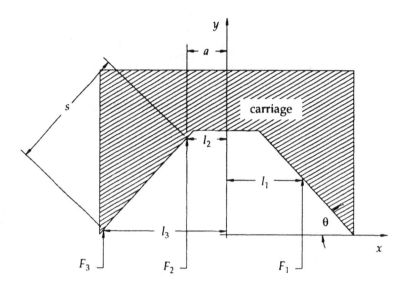

Figure 9.9   Force balance at pads of prismatic way

The forces F_1, F_2 and F_3 represent the reaction to the dead weight of the carriage plus specimen. If there is complete symmetry about the y axis, the total load on each face must be equal for horizontal equilibrium. Rotational equilibrium in the plane of the diagram then makes $F_2 = F_3$. If the coefficients of friction at the pads are equal, there will be no net moment about the y axis, so the centre of friction and the correct drive position also lie on that axis. Assuming, instead, that the load on each face is shared equally by the pads on that face gives the vertical force components as $F/2$, $F/6$ and $2F/6$ respectively where F is the total vertical gravitational force. The normal force is obtained by dividing these by the cosine of the half angle of the prism. During carriage motion at a uniform velocity, the moment about the y axis is

$$M_y = \frac{\mu F}{2\cos(\theta)}\, l_1 - \frac{\mu F}{6\cos(\theta)}\, l_2 - \frac{2\mu F}{6\cos(\theta)}\, l_3 \qquad (9.8)$$

$$\left.\begin{array}{l} l_1 = a + \left(\dfrac{s}{2}\right)\cos(\theta) \\[2ex] l_2 = a \\[2ex] l_3 = a + s\cos(\theta) \end{array}\right\} \qquad (9.9)$$

Substituting Equations (9.9) into (9.8), the total moment about the y axis is given by

$$M_y = -\frac{\mu F s}{12} \qquad (9.10)$$

If s, μ and F are small (for example, 40 mm, 0.1 and 2 N) then so too will be the y axis moment (0.67 N mm). The analysis can be performed in each axis to determine the magnitude of all moments. When, as here, these values turn out to be small, then offsetting the drive axis slightly from the centre of symmetry of the carriage has negligible effect and is often ignored. Care must be taken with these equations because introducing offset forces or torques changes the reaction loads at the pads and invalidates the assumptions used. The analysis procedure outlined in Section 9.1 must then be followed.

Using this way system, stylus profiles of up to 50 mm length can be repeatably measured to within 1 nm at a carriage traverse speed of 0.1 mm s^{-1} with a 25 Hz filter on the stylus output, Lindsey et al., 1988. Almost certainly this does not reveal the full potential of polymeric bearings for sub-nanometre applications. Further research into such way systems is proceeding at the time of writing.

9.5.2   Molecular measuring machine

The molecular measuring machine is currently being developed at the American National Institute of Standards and Technology (NIST) for the manipulation and examination of surfaces with sub-nanometre precision, Evans et al., 1989. It consists of two orthogonal slideways that should enable mapping over a 50 mm by 50 mm region. Surfaces can be examined by several probing methods such as optical, tunnelling, atomic force, etc. Similar to the Nanosurf design, the slideway consists of a simple prismatic bar upon which a carriage is mounted on five kinematic pads. The significant difference between this design and the previous example is a requirement that the height remain unchanged to better than 0.1 nm during a repeat traverse. Necessarily the wear rate must be extremely low. Working on the *general* principle that wear will be related to the hardness of the softest material in the interface, a system comprising two hard metallic surfaces has been chosen. The carriage pads are of sapphire (or, even better, diamond) coated with titanium nitride. Because of its excellent diffusivity, thermodynamic stability and machinability, oxygen free copper is used for the slideway. Thermal expansion effects are reduced by operating the system in high vacuum with temperature stabilized to better than 0.1 mK. The surface is protected by a phosphorous doped electroless nickel coating. Finally, to reduce the friction coefficient of the bearing, both surfaces

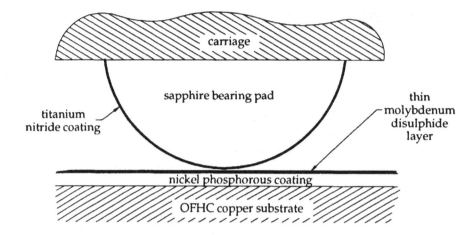

Figure 9.10 Cross-section of a bearing for the molecular measuring machine

are coated with a thin layer of molybdenum disulphide. Using a pin on disc test, wear rates of less than 1 nm per metre traverse have been measured. A schematic cross-section of the bearing is plotted in Figure 9.10.

9.5.3 Super stiff grinding machine slides

The possibility of 'ductile' grinding of brittle materials is attracting much interest as a way of obtaining precise surface shapes with very little sub-surface damage. It is a phenomenon that, as far as we know, occurs with all brittle materials as the depth of cut in the machining process is reduced below a critical value. For some of the more interesting materials such as silicon, germanium, indium phosphide, silica, various zirconates, silicon carbide and silicon nitride this depth is usually below a few hundreds of nanometres. As usual, a process capability of an order of magnitude better precision than this depth is needed, even under cutting load conditions. Unfortunately, the load required for cutting does not reduce with depth and will typically remain relatively constant at a few tens of newtons at the smallest depths of cut. Consequently it necessary to produce a grinding machine having a stiffness in the region of 1 N nm^{-1} (this can be thought of as the stiffness of a 70 mm square by 1 m long or 5 mm cubed steel rods!). To achieve such stiffness in practice, it has proved necessary to use a combination of both ultra stiff bearing design and closed loop compensation. The vertical and horizontal slideways in the pioneering designs of Prof. Miyashita use oil hydrostatic bearings to apply a high constant force to a plain bearing, see Figure 9.3, Miyashita, 1991 (less recent references refer to this as a hydrodynamic bearing which would be the case at high velocity or with viscous lubricating films). To reduce the effects of the high frictional forces which will always induce distortion of the machine, a sophisticated hydraulic load compensator has been introduced into this design. Using such sophisticated techniques, a machine has been constructed that has nanometre level controllability and the desired static stiffness. Its repeatability is at the 100 nm level, Miyashita and Yoshioko, 1982, Rowe *et al.*, 1989.

Boundary lubrication can be used for the bearing pads in this type of application, even though the wear will be large compared to the controlled precision. Maintaining a fixed datum is less important here because the workpiece surface is, in any case, continuously removed by the reciprocating traversal of the specimen against an incrementing grinding head. Consequently, all that is required is that the specimen moves in the same path with a cyclic repeatability of better than the minimum depth of cut. Fortuitously this is well within the capability of a boundary lubricated bearing.

References

Atkins A.G., Omar M.K. and Lancaster J.K., 1984, Wear of polymers, *J. Mater. Sci. Letts.*, **3**, 779-782.

Archard J. F., 1957, Elastic deformation and the laws of friction, *Proc. Roy. Soc.*, **A238**, 190-205

Arnell R.D., Davies P.B., Halling J. and Whomes T.L., 1991, *Tribology: Principles and Design Applications*, Macmillan, London.

Barwell F.T., 1956, *Lubrication of Bearings*, Butterworth, London.

Boenig H.V., 1982, Plasma Science and Technology, Cornell University Press, Ithaca, chpt. 7.

Bowden F. P. and Tabor D., 1964, *The Friction and Lubrication of Solids: Part II*, Clarendon Press, Oxford, Chapters XI & XII.

Briscoe B.J., Evans P.D. and Lancaster J.K., 1987, Single point deformation and abrasion of γ-irradiated poly(tetrafluoroethylene), *J. Phys. D: Appl. Phys.*, **20**, 346-353

Cameron A., 1976, *Basic Lubrication Theory*, 2nd ed., Chichester:Ellis Horwood

Chetwynd D.G. and Siddall G.J., Improving the accuracy of roundness measurement, *J. Phys. E:Sci. Instrum.*, **9**, 537-544

Deacon R. F. and Goodman J. F., 1958, Lubrication by lamellar solids, *Proc. Roy. Soc.*, **A243**, 464-481.

Dowson D., Taheri S. and Wallbridge N. C., 1987, The role of counterface imperfections in the wear of polyethylene, *Wear*, **119**, 277-293

Elwell R. C. and Sternlight B., 1960, Theoretical and experimental analysis of hydrostatic thrust bearings, *Trans. ASME series D; J. Basic Eng.*, **82**(3), 505-512

Evans C.R., Polvani R.S., Scire F.E., Ruff A.W., Hu X.Z. and Teague E.C., 1989, Way systems for the molecular measuring machine, *5th Int. Precision Engineering Seminar*, 18-22 September, Monterey, California

Fort T., 1962, Adsorption and boundary friction on polymer surfaces, *J. Phys. Chem.*, **66**, 1136-1143.

Freeman P., 1962, *Lubrication and Friction*, Sir Isaac Pitman & Sons, London, Chapter 14.

Fuller D.D., 1984, *Theory and Practice of Lubrication for Engineers*, 2nd ed., New York:John Wiley & Sons

Gross W.A., 1962, *Gas Film Lubrication*, John Wiley and Sons, Inc., London.

Hornbogen E. and Karsch U.A., 1983, Frictional wear of polytetrafluoroethylene, *J. Mater. Sci. Letts.*, **2**, 777-780.

Johnson K.L., 1987, *Contact Mechanics*, Cambridge University Press, Chapter 13.

Lancaster J.K., 1984, The lubricated wear of polymers, 11th Leeds-Lyons Symposium on *Mixed Lubrication and Lubricated Wear*, Leeds 4 -7 Sept. 1984.

Laub J. H., 1960, Hydrostatic gas bearings, *Trans. ASME series D; J. Basic Eng.*, **82**(2), 276-286

Lees W.A., 1989, *Adhesives and the Engineer*, Mechanical Engineering Publications Ltd., London.

Lhymn C., 1986, Microscopy study of the frictional wear of polytetrafluoroethylene, *Wear*, **107**, 95-105.

Lindsey K., Smith S. T. and Robbie C. J., 1988, Sub-nanometre surface texture and profile measurement with Nanosurf 2, *Annals of the CIRP*, **37**(1), 519-522.

Mackinson K.R. and Tabor D., 1964, The friction and wear of polytetrafluoroethylene, *Proc. Roy. Soc.(Lond.)*, **A281**, 49-61.

McLaren K.G. and Tabor D., 1965, The friction and deformation properties of irradiated polytetrafluoroethylene (PTFE), *Wear*, **8**, 3-7.

Miyashita M. and Yoshioka J., 1982, Development of ultra-precision machine tools for micro-cutting of brittle materials, *Bull. Japan Soc. of Prec. Engg.*, **16**(1), 43-50.

Muijderman E.A., 1966, *Spiral Groove Bearings*, Cleaver-Hume Press Ltd., London.

O'Leary K. and Geil P.H., 1967, Polytetrafluoroethylene fibril structure, *J. Appl. Phys.*, **8**, 4169-4181.

Pauling L., 1945, *The Nature of the Chemical Bond*, Cornell University Press, Ithaca, 106-108.

Reynolds O., 1886, On the theory of lubrication and its application to Mr Beauchamp Tower's experiments, including an experimental determination of the viscosity of olive oil. *Phil. Trans. Roy. Soc. London*, **177**, 157-234

Rowe W. B., Miyashita M. and Koenig W., 1989, Centreless grinding research and its application to advanced manufacturing technology, *Annals CIRP*, **38**(2), 617-625

Rowlands H. A., 1902, *The Physical Papers of H. A. Rowlands*, The Johns Hopkins Press, Baltimore, 691-706.

Rubenstein C., 1961, Lubrication of polymers, *J. Appl. Phys.*, **32** (8), 1445-1450.

Savage R. H., 1948, Graphite Lubrication, *J. Appl. Phys.*, **19**(2), 1-10.

Senior J.M. and West G.H., 1971, Interaction between lubricants and plastic bearing surfaces, *Wear*, **18**, 311-323.

Slocum A.H., 1992, *Precision Machine Design*, Prentice Hall, NJ.

Sneck H. J., 1968, A survey of gas lubricated journal bearings, *Trans. ASME; J. Lubr. Eng.*, **90**, 804-809

Steijn R.P., 1968, The sliding surface of polytetrafluoroethylene; An investigation with the electron microscope, *Wear*, **12**, 193-212.

Tabor D., 1957/58, Friction, lubrication and wear of synthetic fibres, *Wear*, **1**, 5-24.

Tabor D., 1982, The role of surface and intermolecular forces in thin film lubrication, in Georges J.M. (ed.) *Microscopic Aspects of Adhesion and Lubrication*, Elsevier, Amsterdam.

Tanaka K., Uchiyama Y. and Toyooka S., 1973, The mechanism of wear of polytetrafluoroethylene, *Wear*, **23**, 153-172.

10 THE DYNAMICS OF INSTRUMENTATION MECHANISMS

Precision mechanical instruments are rarely highly stressed through dynamical effects and so the methods of statics can be used for most of their design. They are very susceptible to small scale oscillatory behaviour that can limit their ability to function accurately in the presence of unwanted disturbances or to follow a signal. Consequently system dynamics is discussed here in the context of response and vibration isolation. The first part of the chapter examines basic analytical techniques, concentrating on the readily generalized combination of Lagrangian mechanics and transfer function (frequency response) descriptions. The dynamic response of linear second order, single degree of freedom systems is used to illustrate characteristic behaviours. Particular emphasis is placed on viscous damping in the system, expressed by the critical damping factor, ξ, and the related quality factor, Q. Connecting these elements in series results in higher order systems which are briefly considered because of their importance in vibration isolation. The practical measurement of damping is then examined, particularly the 'hammer test' and logarithmic decrement method. This is followed by a discussion of parameter selection, especially damping coefficients, for a variety of applications, mainly noise and vibration isolation. Finally, there is a review of methods of damping suitable for small amplitude applications, including acoustic chambers, rubber and air suspensions and high damping factor materials. Case studies cover aspects of scanning tip microscope and gravity wave antenna designs.

10.1 The simple oscillator

This chapter takes a brief look at the dynamic behaviour of instrument mechanisms, particularly as it limits their performance. Response to sudden changes of input and sensitivity to vibration are the main concerns. Resonance in individual elements and complete structures is of importance to both. A common experience of using instruments to the limits of precision is that of working through many long nights, because external disturbances from fellow workers and other machinery is lower then – perhaps an improved understanding of dynamics at this level can literally put us in the light. Resonance is everywhere in nature and so familiar that a general description is unnecessary. However, it is useful to review the simplest mechanical oscillator systems in order to confirm the physical assumptions that are made also in more complex analyses.

To obtain simple, tractable models of the response of a system to a dynamically varying input, or demand, signal, we attempt to describe it by a set of linear equations. The simplest case is where the output is linearly proportional to the input for all values of demand. However, almost invariably energy (power) is to be transferred from one element to another and energy storage in the mechanisms will cause transmission delays. The system response can then only be adequately described by differential equations. For systems described by a set of *linear* differential equations, the amplitude of the system response to a *steady state* demand is linearly proportional to its frequency content and its associated delay, or phase lag. If the signal consists of a number of frequencies at the input, the principle of superposition applies and the amplitudes and phase lags of each frequency can be added either vectorially or trigonometrically. Alternatively if the amplitude, frequency and phase of an input is known, the amplitude and phase of the output can be predicted. The frequency remains *unchanged* if the system is linear.

All mechanical elements have mass and elasticity. In lumped parameter modelling we simplify by assuming that a rigid inertia is fixed to a rigid base through a relatively compliant element of negligible mass, as in the spring-mass system of Figure 10.1. The crucial feature of such systems is that both the mass and spring can store energy, respectively as kinetic energy and strain (potential) energy. *Any* mechanical system that combines potential and

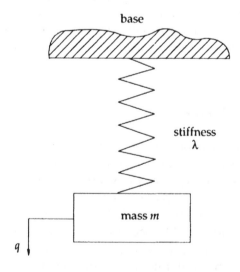

Figure 10.1 A simple single degree of freedom spring/mass system

kinetic energy storage elements can exchange a fixed amount of energy backwards and forwards between the two forms without violating the laws of conservation. This is the fundamental source of oscillation. For example, if the spring-mass system of Figure 10.1 has mechanical energy, E, at some instant and no transfer out of the system is allowed, then its total energy will be at all times the sum of kinetic and potential energies:

$$E = \frac{1}{2}m\dot{q}^2 + \frac{1}{2}\lambda q^2 \qquad (10.1)$$

where m is the lumped mass, λ the spring stiffness and q the coordinate of rectilinear motion. For this conservative system, we find, by differentiating with respect to time and noting that the velocity could only be zero at all times in the trivial case, that

$$m\ddot{q} + \lambda q = 0 \qquad (10.2)$$

Equation 10.2 may be rewritten as the familiar linear, second order harmonic oscillator equation

$$\ddot{q} + \omega_n^2 q = 0 \qquad (10.3)$$

where $\omega_n = (\lambda/m)^{1/2}$ is its natural, or resonant, frequency. At instants of zero velocity, the system holds only potential energy in the form of strain energy and the spring will be at its maximum distortion, so

$$E = \frac{1}{2}\lambda q_{max}^2 \qquad (10.4)$$

The maximum amplitude of motion for a given input energy, E, is

$$q_{max} = \sqrt{2E/\lambda} \qquad (10.5)$$

Equation 10.5 indicates that, for a given input energy and in the absence of significant energy dissipation mechanisms, the amplitude of motion can be reduced only by increasing the stiffness.

All practical systems will be excited by internal (motors, gearboxes and crankshafts) and external (adjacent machinery, traffic, people) sources of vibration and by airborne noise. Thus there must be a means of dissipation in every instrument to prevent it slowly absorbing energy and building up oscillations of an unacceptable magnitude. The principle energy dissipation characteristics are viscous, hysteretic and coulomb damping. The first two are most commonly used and also have similar transient characteristics. Thus we analyse only system responses where energy is dissipated by a viscous damper, that is one providing a resisting force directly proportional to the velocity of motion.

10.2 Response of a second order spring/mass/viscous damper system

In Figure 10.2 a viscous damper having a dissipative constant b (dimensionally of units N s m^{-1}) has been attached in parallel with the spring of the basic oscillator of Figure 10.1. As discussed in Chapter 2, Lagrange's method applies to conservative systems but it may be extended, Landau and Lifshitz, 1960, by introducing a viscous dissipation function, $\mathcal{D}$, of the form

$$\mathcal{D} = \frac{1}{2} b \dot{q}_i^2 \tag{10.6}$$

and adding it into 2.31 to give

$$\sum_{i=1}^{n} \left\{ \frac{\mathrm{d}}{\mathrm{d}t} \left(\frac{\partial \mathcal{L}}{\partial \dot{q}_i} \right) - \frac{\partial \mathcal{L}}{\partial q_i} + \frac{\partial \mathcal{D}}{\partial \dot{q}_i} = Q_i \right\} \tag{10.7}$$

where b is the viscous damping coefficient, $\mathcal{L} = (T - U)$ is the Lagrangian, the difference of the kinetic and potential energies, and Q_i the applied force in the direction of the ith coordinate q_i. Consistent with our earlier expressed views that Lagrangian methods are appropriate because they allow an easy method

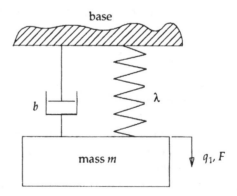

Figure 10.2   A spring/mass/damper system with forced excitation

for extension to systems of considerable complexity, we shall use this form in the analysis of simple systems in the following sections.

10.2.1 Standard forms of the equation of motion

In the system of Figure 10.2 Q_i is simply F and the energy terms are the same as in Equation 10.1, so Equation 10.7 readily gives the motion of this single degree of freedom system as

$$m\ddot{q}_1 + b\dot{q}_1 + \lambda q = F \tag{10.8}$$

Its *transient* behaviour is of the general form

$$q_1(t) = Ae^{\alpha t} \tag{10.9}$$

The second order system will be quadratic in α, which is found by direct substitution into the homogeneous form of Equation 10.8, giving

$$m\alpha^2 + b\alpha + \lambda = 0 \tag{10.10}$$

This characteristic equation has two roots at

$$\alpha = \frac{-b \pm \sqrt{b^2 - 4m\lambda}}{2m} \tag{10.11}$$

or

$$\alpha = -\xi\omega_n \pm \omega_n\sqrt{\xi^2 - 1} \tag{10.12}$$

where ξ is the damping factor (or coefficient), defined as

$$\xi = \left(\frac{b^2}{4m\lambda}\right)^{1/2} \tag{10.13}$$

The general transient solution is a superposition of Equation 10.9 for each independent value of α

$$q_1(t) = Ae^{\left(-\xi\omega_n + \omega_n(\xi^2 - 1)^{1/2}\right)t} + Be^{\left(-\xi\omega_n - \omega_n(\xi^2 - 1)^{1/2}\right)t}$$

$$= e^{(-\xi\omega_n t)}\left(Ae^{\left(\omega_n(\xi^2 - 1)^{1/2} t\right)} + Be^{\left(-\omega_n(\xi^2 - 1)^{1/2} t\right)}\right) \tag{10.14}$$

Depending on the value of ξ, Equation 10.14 takes on one of three different characteristics. If $\xi > 1$ the square root term in Equation 10.14 is real and positive, there can be no oscillatory motion and the response is the sum of two exponential decays of different time constants. This is known as an *overdamped* system. For $\xi < 1$ the square root expression is complex, indicating oscillatory behaviour, and Equation 10.14 becomes

$$q_1(t) \; = \; e^{(-\xi\omega_n t)}\left(Ae^{(i\omega_d t)} \; + \; Be^{(-i\omega_d t)} \right) \tag{10.15}$$

The frequency of the oscillatory term in Equation 10.15 is reduced from that of the undamped natural frequency and is given by

$$\omega_d \; = \; \omega_n\sqrt{1 - \xi^2} \tag{10.16}$$

Using De Moivre's theorem, Equation 10.15 can be expanded and simplified to give

$$q_1(t) \; = \; e^{(-\xi\omega_n t)}(C\cos(\omega_d t) \; + \; D\sin(\omega_d t)) \tag{10.17}$$

where the C and D are constants satisfying the two boundary conditions necessary for the complete description of a second order system. Such *underdamped* systems display transients consisting of exponentially decaying sinusoids. The time constant of the decay envelope is $1/\xi\omega_n$. Technically the transient is not periodic, but it displays maxima at fixed time intervals, often called its pseudo-period, corresponding to a damped natural frequency, ω_d.

The third solution to Equation 10.14 involves the singular instance of repeated roots when ξ is unity. Behaviour is intermediate between oscillatory and non-oscillatory and such a system is described as *critically damped*. It is in some sense an ideal form, but in practice very difficult to achieve and more so to maintain throughout the lifespan of an instrument. To avoid the long settling times associated with overdamped systems, most designs permit some oscillatory behaviour with ξ set in the region 0.05 - 0.7.

Equations 10.14 and 10.15 indicate the initial behaviour of the system. Its steady state behaviour is described by the particular integral of Equation 10.8, which, using Equation 10.13, can be rearranged as

$$\ddot{q}_1 \; + \; 2\xi\omega_n \dot{q}_1 \; + \; \omega_n^2 q_1 \; = \; \frac{F(t)}{m} \tag{10.18}$$

With this set of equations it is a relatively straightforward exercise to determine the response of second order systems to a range of simple input force functions.

10.2.2 Response to a step change of force

The step response of second order, or higher order, systems is a common way of summarizing their transient dynamic behaviour. Consider the case where the mass in Figure 10.2 has been first displaced a distance δ by a stationary applied force F, held and then released at time $t = 0$. It is intuitively obvious that the mass will eventually settle in its original position, so the steady state solution is of zero motion. Because at $t = 0$, $q_i(t) = F/\lambda$, the constant C in Equation 10.15 must equal F/λ. The constant D is found, since the initial velocity is zero, by differentiating Equation 10.17 at $t = 0$ to give

$$\dot{q}_1(0) = 0 = D\omega_d \cos(\omega_d 0) - \frac{F}{\lambda} \xi \omega_n \cos(\omega_d 0) \qquad (10.19)$$

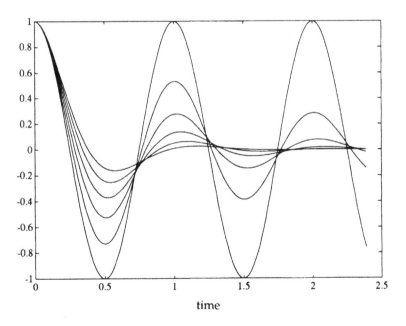

Figure 10.3 Step response of the spring/mass/damper system of figure 10.2 against time for a system of natural frequency of $1/2\pi$. Values for ξ are $0, 0.1, 0.2, 0.3, 0.4$ and 0.5 respectively

Therefore the complete equation of motion for a step input to this system is

$$q_1(t) = \frac{F}{\lambda} e^{(-\xi\omega_n t)} \left(\cos(\omega_d t) + \frac{\xi}{\sqrt{1-\xi^2}} \sin(\omega_d t) \right) \tag{10.20}$$

The step responses for various values of damping factor are plotted in Figure 10.3. Note how an increase in the damping factor not only increases the decay rate of the envelope but also, through its effect on the damped natural frequency, increases the time between successive zero crossings.

Design specifications often use the criterion of settling time, defined as the time required for the system to settle within a certain percentage of the input. For second order systems the requirement typically specifies a maximum delay before the ouput remains within 2% of its final value after a step input change, which takes a duration of four time constants ($4\tau = 4/\xi\omega_n$). The system response is improved by increasing both the fundamental resonant frequency and the damping factor, providing the latter remains less than unity. However, this is true only in the steady state. If the system needs to respond to frequent changes in the demand the reduction in the natural frequency may reasonably be expected to influence the accuracy with which the output follows the input. Some account of this might be taken by using the damped natural frequency, $\omega_n(1 - \xi^2)^{\frac{1}{2}}$, rather than ω_n in the estimate of the time constant. Then differentiating with respect to the damping factor gives

$$\frac{\partial\tau}{\partial\xi} = \frac{1}{\xi^2\omega_n} \frac{(2\xi^2 - 1)}{(1-\xi^2)^{\frac{3}{2}}} \tag{10.21}$$

The time constant is minimum, perhaps implying an optimal response, when the numerator of Equation 10.21 is zero, that is when ξ is about 0.7. Other arguments that weigh the time to achieve the first peak overshoot after a step input against the percentage overshoot indicate optimal damping factors in the region 0.3 to 0.6, Dorf, 1980. Section 10.5 gives some further discussion on the choice of appropriate damping ratios.

10.2.3 Steady state response to a harmonically varying input force

Many instrument mechanisms and other machines involve a combination of springs and masses with some form of inherent (frictional) damping. Their behaviour in the presence of continuous, unwanted but inevitable, external disturbances is of great interest. Additionally, to reduce the influence of external and internal perturbations, whole instruments are mounted on vibration isolators consisting of compliant supports in parallel with energy dissipation devices. In general the simplest device for energy dissipation is the inclusion of a viscous damper. It is likely that in either case the input forces will be of a random or unpredictable nature. There are some sophisticated methods within random process theory by which limited progress can be made towards predicting system response under such conditions, see, for example, Whitehouse, 1988. However, commonly, as here, the difficulty in analysis is sidestepped by noting that any signal of finite length can be closely approximated by a Fourier series. Since the response of a system described by linear differential equations is linearly related to the phase, amplitude and frequency of the input, we may consider in turn the effect of each term in the Fourier series approximation and superpose the results.

Again to provide easy generalization to more complicated systems, it is convenient to express a cosinusiodal input force of magnitude F as the real part of the complex quantity $Fe^{(i\omega t)}$, where $i = \sqrt{-1}$. The system is then modelled as a transfer function between this force and the output, q_1, such that

$$q_1(t) = Re\left[H(i\omega) Fe^{(i\omega t)}\right] \tag{10.22}$$

where $H(i\omega)$ is the frequency response, a complex quantity having the general properties

$$H(i\omega) = A(\omega) + iB(\omega)$$

$$|H| = \sqrt{A^2 + B^2} \tag{10.23}$$

$$arg(H) = \tan^{-1}(B\!/\!A) = \varphi$$

where A and B are real constants at any given frequency. We now derive H and so show that the transfer function concept is summarized by the flow diagram:

$$F\cos\omega t = Re(Fe^{i\omega t}) \longrightarrow \boxed{H(i\omega)} \longrightarrow F|H|\cos(\omega t - \varphi)$$

To determine the frequency response, the output form of Equation 10.22 is substituted into the equation of motion, 10.18, to give

$$(-\omega^2 + 2\xi\omega_n\, i\omega + \omega_n^2)\, FH(i\omega)\, e^{(i\omega t)} = \frac{F}{m} e^{(i\omega t)} \tag{10.24}$$

from which

$$H(i\omega) \ = \ \frac{1/m}{-\omega^2 + 2\xi\omega_n i\omega + \omega_n^2} \ = \ \frac{1/\lambda}{\left(1 - \dfrac{\omega^2}{\omega_n^2}\right) + 2i\,\xi\,\dfrac{\omega}{\omega_n}} \tag{10.25}$$

The transfer function of Equation 10.25 can be brought into the form of Equation 10.23 by multiplying its numerator and denominator by the complex conjugate of the latter. Then substituting it into Equation 10.22 gives

$$q_1(t) \ = \ Re\left\{\frac{F}{\lambda}\ \frac{1 - \dfrac{\omega^2}{\omega_n^2} - 2i\,\xi\,\dfrac{\omega}{\omega_n}}{\left(1 - \dfrac{\omega^2}{\omega_n^2}\right)^2 + 4\,\xi^2\,\dfrac{\omega^2}{\omega_n^2}}\ e^{(i\omega t)}\right\} \tag{10.26a}$$

$$q_1(t) \ = \ \frac{F}{\lambda}\ \frac{\left(1 - \dfrac{\omega^2}{\omega_n^2}\right)\cos(\omega t) + 2\,\xi\,\dfrac{\omega}{\omega_n}\sin(\omega t)}{\left(1 - \dfrac{\omega^2}{\omega_n^2}\right)^2 + 4\,\xi^2\,\dfrac{\omega^2}{\omega_n^2}} \tag{10.26b}$$

A standard trigonometric relationship shows that the addition of sine and cosine terms of the same argument can be re-expressed as a single cosine of that argument with an additional phase *lag*, so Equation 10.26b may also be written

$$q_1(t) \ = \ \frac{(F/\lambda)\cos(\omega t - \varphi)}{\left(\left(1 - \dfrac{\omega^2}{\omega_n^2}\right)^2 + 4\,\xi^2\,\dfrac{\omega^2}{\omega_n^2}\right)^{1/2}} \tag{10.27a}$$

$$q_1(t) \ = \ F|H|\cos(\omega t - \varphi) \tag{10.27b}$$

where the magnitude of the transfer function is seen by direct comparison of the Equations 10.27 and the phase lag is

$$\varphi \ = \ \tan^{-1}\frac{2\,\xi\,\dfrac{\omega}{\omega_n}}{\left(1 - \dfrac{\omega^2}{\omega_n^2}\right)} \tag{10.28}$$

While Equations 10.27 provide a strictly correct definition of the transfer function between a force input and a displacement output, a non-dimensional form is commonly preferred. From Equation 10.27a, the maximum magnitude of the output under *dynamic* loading, X, is related to the magnitude of deflection under an equivalent *static* load F/λ by

$$\frac{X}{F/\lambda} = \frac{1}{\left(\left(1 - \frac{\omega^2}{\omega_n^2}\right)^2 + 4\,\xi^2\,\frac{\omega^2}{\omega_n^2}\right)^{1/2}} \qquad (10.29)$$

Figure 10.4 shows this amplitude gain, the dynamic amplification, over input frequencies normalized to the undamped natural frequency, ω_n, for a range of damping factors. The shape of the transfer function is characteristic of all second order systems and, also, closely approximates the response of many

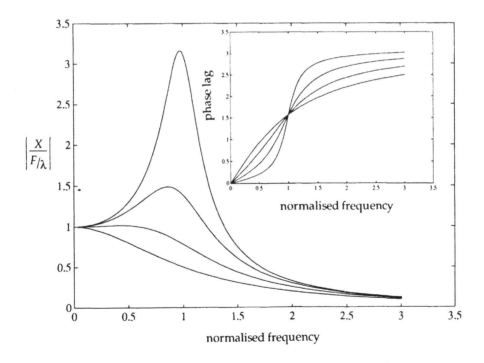

Figure 10.4 Amplitude response of a damped spring mass system subject to a harmonic excitation with damping factors of $\xi=0.16$, 0.36, 0.64 and 1.0; phase response (inset).

295

higher order systems up to and beyond their first natural frequency. The 'gain' is near to unity at low frequencies, well below the undamped natural frequency, and gradually increases up to a maximum at the frequency

$$\omega = \omega_n \sqrt{1 - 2\xi^2} \qquad (10.30)$$

Note that in the presence of damping the peak amplification occurs at a frequency lower than ω_n. Beyond this point the response of the second order system continuously decreases. Dynamic amplification may be curtailed or prevented by increasing the damping factor, but doing so also introduces a considerable phase lag. This may be unacceptable. Choosing a suitable degree of damping is very much application related, see also Section 10.5.

When systems are designed for operation at resonance (as in tuned electrical circuits and mechanical vibrators) it is common to refer to the peak dynamic amplification as the quality factor or Q value. For $\xi \ll 1$, as would be the case in such applications, the peak occurs very close to the undamped natural frequency and has magnitude

$$Q \approx \frac{1}{2\xi} \qquad (10.31)$$

10.2.4 Response to a harmonic displacement: Seismic isolation

Varying forces are not the only source of input disturbance that can affect a second order system. Displacements of the nominally stationary base cause accelerations of the sprung mass and so tend to excite resonant behaviour. Typically such disturbance is transmitted through the floor from nearby traffic, machinery, people or seismic action. As argued in the previous section, we can gain a reasonable understanding of the response to such disturbances by analyzing the special case of excitation by independent sinusoidal motions at different frequencies. In this instance, the system model is that of Figure 10.2 with the base subjected to a harmonic displacement $y(t)$ of the form

$$y(t) = \mathrm{Re}\,(Ye^{(i\omega t)}) \qquad (10.32)$$

Newton's laws indicate that, whereas velocity and displacement are expressed in relative coordinates, accelerations must be referred to the conceptual inertial frame. Hence, if we allow the base to move and apply no external force, Equation 10.8 becomes

$$m\ddot{q}_1 + b(\dot{q}_1 - \dot{y}) + \lambda(q_1 - y) = 0 \qquad (10.33)$$

or, by rearranging and introducing our standard parameters,

$$\ddot{q}_1 + 2\xi\omega_n\dot{q}_1 + \omega_n^2 q_1 = 2\xi\omega_n\dot{y} + \omega_n^2 y \qquad (10.34)$$

We now adopt the same transfer function approach as used for force excitation. The process is a little more obvious this time, requiring simply that Equation 10.30 be substituted into Equation 10.34, to express the response as

$$q_1(t) = \mathrm{Re}\left\{\frac{1 + i2\xi\dfrac{\omega}{\omega_n}}{1 - \left(\dfrac{\omega}{\omega_n}\right)^2 + i2\xi\dfrac{\omega}{\omega_n}} Ye^{i\omega t}\right\} \qquad (10.35)$$

Again, the real part of this Equation yields superposed sine and cosine terms which may be combined into a single lagging cosinusoid. The dynamic response of the table to motion of the foundation is then

$$q_1(t) = Y\left[\frac{1 + \left(2\xi\dfrac{\omega}{\omega_n}\right)^2}{\left[1 - \left(\dfrac{\omega}{\omega_n}\right)^2\right]^2 + \left(2\xi\dfrac{\omega}{\omega_n}\right)^2}\right]^{1/2} \cos(\omega t - \varphi) \qquad (10.36a)$$

$$= X\cos(\omega t - \varphi) \qquad (10.36b)$$

where the magnitude of the transfer function is defined by direct comparison of the Equations 10.36 and the phase angle is

$$\varphi = \tan^{-1}\left[\frac{2\xi\left(\dfrac{\omega}{\omega_n}\right)^3}{1 - \left(\dfrac{\omega}{\omega_n}\right)^2 + \left(2\xi\dfrac{\omega}{\omega_n}\right)^2}\right] \qquad (10.37)$$

This time, the strictly defined transfer function is non-dimensional and indicates the ratio of the peak output displacement to the peak input displacement. It is usually referred to as the *displacement transmissibility* or *receptance* and is plotted against dimensionless frequency in Figure 10.5. An important aspect of this curve is that the dynamic amplification never falls

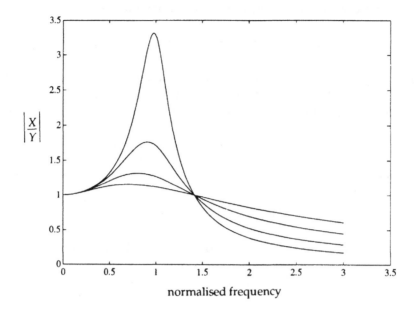

Figure 10.5 Transmissibility of a single degree of freedom, damped spring and mass system ξ = 0.16, 0.36, 0.64 & 1.0

below unity until the input frequency increases above $\sqrt{2}\ \omega_n$. Above this crossover, the attenuation is dependent upon the damping, with smaller damping factors giving more rapid attenuation as the frequency increases. Thus, for example, the designer of a seismic isolator must compromise between a high peak dynamic amplification of unexpected, low frequencies and a relatively slow roll-off in the design frequency range. As always, damping reduces the frequency of the peak amplitude to somewhat below the natural frequency.

10.3 Harmonic response of series spring-mass-damper systems

So far we have concentrated on the dynamics of simple, single degree of freedom mechanisms to illustrate typical behaviour patterns. Practical instruments almost always involve the combination of interconnected elements that display second order (or higher) dynamical characteristics. Analyzing such systems is not particularly difficult, but it is always tedious and, therefore, error-prone. Lagrange's equations, as explained in Chapter 2 and extended at Equation 10.7, come into their own for determining the equations of motion of multi-degree of freedom systems because they are highly systematized. For present purposes we analyze only a two-stage system, which is sufficient to illustrate some general features without involving too much mathematical complexity.

10.3.1 Response of two stage seismic isolator

We illustrate the generalization of the approach developed in Section 10.2 by considering the typical case of an instrument mounted on an anti-vibration table. It may be modelled as shown in Figure 10.6. The table upon which the instrument is mounted is isolated by a spring/damper support designed to attenuate the amplitude of displacements of the foundation. The instrument is mounted rigidly onto the table, but itself acts as a second order system. For example, a measuring head might be mounted onto an adjustable column to enable its vertical positioning relative to a specimen. The column may be less rigid than most of the other components and so introduce the possibility of relatively large vibrations between the sensing head and base. These would be

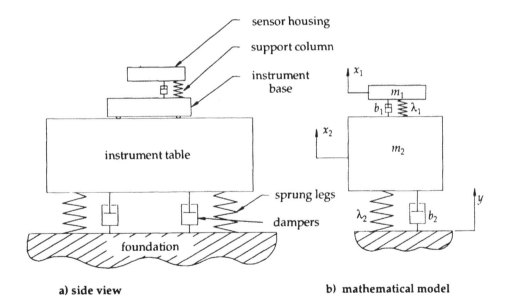

a) side view   b) mathematical model

Figure 10.6 Schematic representation of a laboratory instrument mounted on an anti-vibration table; a) schematic model, b) simplified model ignoring torsional freedoms

measured as noise by the sensor. To avoid lengthy equations, the simplified model shown in Figure 10.6b will be used. Including inertial terms to model the mass of the springs and finite torsional stiffnesses and moments of inertia will simply add to the number of degrees of freedom. The equations of motion, although just as simple to derive, then become increasingly cumbersome and are best solved numerically.

In the model of Figure 10.6b the linear motions of the sensor and table relative to a fixed inertial frame are represented by the coordinates x_1 and x_2. The displacement of the ground is similarly given by y. Noting that the spring and damper displacements are differential between these coordinates, the kinetic energy, potential energy and dissipation functions of this system are

$$T = \tfrac{1}{2}(m_1\,\dot{x}_1^2 + m_2\,\dot{x}_2^2)$$

$$U = \tfrac{1}{2}(\lambda_1(x_1 - x_2)^2 + \lambda_2(x_2 - y)^2) \qquad (10.38)$$

$$D = \tfrac{1}{2}(b_1(\dot{x}_1 - \dot{x}_2)^2 + b_2(\dot{x}_2 - \dot{y})^2)$$

Substituting these functions into the extended Lagrange formula, Equation 10.7, with no forcing function yields the system equations of motion

$$m_1\ddot{x}_1 + b_1(\dot{x}_1 - \dot{x}_2) + \lambda_1(x_1 - x_2) = 0$$

$$m_2\ddot{x}_2 + b_1(\dot{x}_2 - \dot{x}_1) + b_2(\dot{x}_2 - \dot{y}) + \lambda_1(x_2 - x_1) + \lambda_2(x_2 - y) = 0 \qquad (10.39)$$

Again, we examine the steady state response to harmonic excitation through the transfer function approach used in Section 10.2. The motion of the foundation is taken to be of the form $y(t) = \mathrm{Re}\,(Ae^{i\omega t})$. The resulting motion of the coordinates x_1 and x_2 is then

$$x_1 = \mathrm{Re}\,[\,H_{x_1 y}(i\omega)\,A\,e^{i\omega t}\,]$$

$$x_2 = \mathrm{Re}\,[\,H_{x_2 y}(i\omega)\,A\,e^{i\omega t}\,] \qquad (10.40)$$

where $H_{\alpha\beta}(i\omega)$ is the linear, complex frequency response or receptance of the coordinate α to an input β.

Direct substitution of Equations 10.40 into Equations 10.39 yields, after algebraic rearrangement, the transfer functions relating the output motion of the sensor head and the table to the input foundation motion

$$H_{x_1 y}(i\omega) = \frac{(\lambda_2 + i\omega b_2)(\lambda_1 + i\omega b_1)}{\omega^4 m_1 m_2 - i\omega^3 [m_1(b_1 + b_2) + m_2 b_1]}$$

$$\cdots \frac{1}{+ i\omega[b_1 \lambda_2 + b_2 \lambda_1]}$$

$$\cdots \frac{1}{- \omega^2 [\lambda_1(m_1 + m_2) + \lambda_2 m_1 + b_1 b_2] + \lambda_1 \lambda_2} \qquad (10.41a)$$

$$H_{x_2 y}(i\omega) = H_{x_1 y}(i\omega)(i\omega b_1 + \lambda_1 - \omega^2 m_1) / (i\omega b_1 + \lambda_1) \qquad (10.41b)$$

If the damping becomes negligibly small, the imaginary terms disappear to leave the real transfer function

$$H_{x_1 y}(i\omega) = \frac{\lambda_1 \lambda_2}{\omega^4 m_1 m_2 - \omega^2 [\lambda_1(m_1 + m_2) + \lambda_2 m_1] + \lambda_1 \lambda_2} \qquad (10.42)$$

This has two poles that can be readily found because the denominator of Equation 10.42 is a quadratic in ω^2. The solution values correspond to the two natural frequencies of the undamped system.

Even for this two degree of freedom system, the full transfer functions are too complicated to give much help to intuition. Following, still, the methods used in Section 10.2, Equations 10.41a and 10.39 combine to give the steady state frequency and phase response of the sensor head as

$$x_1 = \left(\frac{C^2 + D^2}{A^2 + B^2} \right)^{1/2} \cos(\omega t - \varphi)$$

$$\varphi = \tan^{-1}\left(\frac{CB - DA}{AC + BD} \right) \qquad (10.43)$$

$$A = \omega^4 m_1 m_2 - \omega^2 [\lambda_1(m_1 + m_2) + \lambda_2 m_1 + b_1 b_2] + \lambda_1 \lambda_2$$

$$B = \omega[b_1 \lambda_2 + b_2 \lambda_1] - \omega^3 [m_2 b_1 + m_1 (b_1 + b_2)]$$

$$C = \lambda_1 \lambda_2 - \omega^2 b_1 b_2 \qquad (10.44)$$

$$D = \omega[b_1 \lambda_2 + b_2 \lambda_1]$$

This frequency response is plotted in Figure 10.7. The equivalent form for x_2 is not given explicitly but is shown in figure 10.7c. In this particular example, the

301

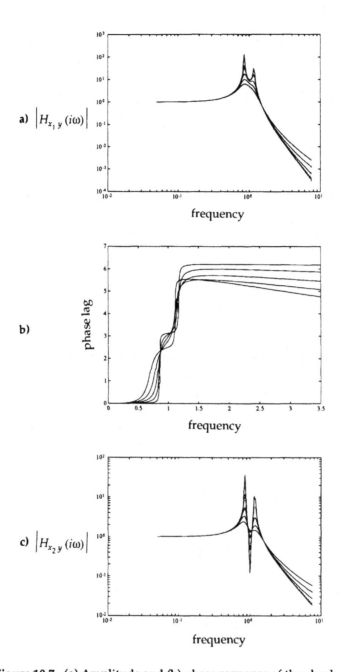

Figure 10.7 (a) Amplitude and (b) phase response of the absolute motion of m_1 and (c) amplitude response of the absolute motion of m_2 for the second order system of Figure 10.6b. Values of system elements are $m_1=1$, $m_2=10$, $\lambda_1=1$, $\lambda_2=10$. Damping values are $b_1=0.012$, 0.048, 0.11, 0.19, 0.30 and $b_2=0.16$, 0.64, 1.44, 2.56, 4.00 respectively.

dynamic measurement error of the instrument is simply the differential displacement between the two masses. This transfer function is

$$H_{x_1 x_2}(i\omega) \;=\; H_{x_1 y}(i\omega) \;-\; H_{x_2 y}(i\omega) \tag{10.45}$$

For the parameter values chosen, the error is considerably greater than the motion of the foundation over a large frequency range. This is because the difference between the two natural frequencies of the system is rather small. To achieve reasonably effective attenuation of foundation noise, it is desirable that the natural frequency of the instrument be at least an order of magnitude greater than that of the isolation stage. Sufficient damping must also be present to prevent excessive amplification near to the natural frequencies. Figure 10.8 shows the frequency response of the same system with parameters selected to place the undamped natural frequencies approximately one order of magnitude apart. The effective measurement error, Figure 10.8c, is now much smaller than the excitation amplitude, by approximately two orders of magnitude over much of the range and to a few percent near the natural frequencies.

10.3.2 Multi-stage isolation

The previous section illustrates the complexity of the analytic solutions of even the simplest multi-freedom systems. There is little point in deriving general transfer functions for others here, but Equations 10.41 indicate a pattern of behaviour from which a number of useful inferences about more complex systems can be obtained. Firstly, in the absence of damping the denominator of the transfer function contains zeros, corresponding to the frequency of displacement resonances. However, as the frequency increases above that of the highest resonance there is a steady attenuation at a rate that asymptotically approaches inverse proportionality to the fourth power of the frequency. As shown in Section 10.2.4, above a certain value the dynamic amplification reduces to less than unity with the lower values of damping factor causing a faster reduction rate. Thus we have the notion that mounting several isolation stages in series will result in an enhanced attenuation rate at sufficiently high frequencies. 10.8 illustrates that the attenuation is enhanced with reduced damping at frequencies above

$$\omega^2 \;=\; \lambda_1\!\left(\frac{1}{m_1} + \frac{1}{m_2}\right) + \frac{\lambda_2}{m_2} \tag{10.46}$$

Clearly, when the two stiffnesses and masses are the same the cross-over frequency is simply $\omega^2 = 3\lambda/m$.

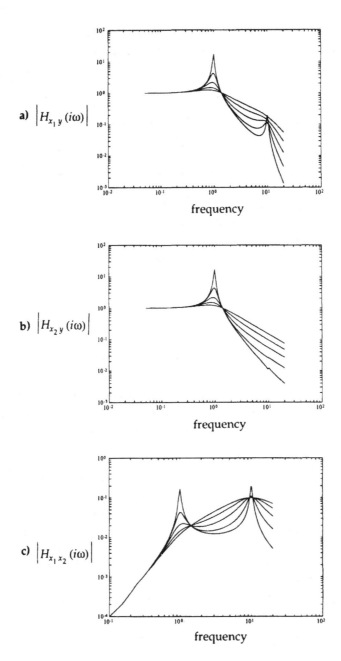

Figure 10.8 Amplitude of absolute motion of masses a) m_1 and b)m_2, c) the relative amplitude response between masses 1 and 2 for the second order system of Figure 10.7. Values of system elements are m_1=0.1, m_2=10, λ_1=10, λ_2=10. Damping values are b_1=0.06, 0.09, 0.17, 0.38, 1.5 and b_2=0.60, 0.94, 1.67, 3.80, 15.0 respectively.

With more series spring/mass elements this analysis becomes rather cumbersome and it is appropriate to employ more advanced computational techniques such as those outlined by Newland, 1990. Such techniques give the cross-over frequencies for three and four mass systems as $\omega^2 = 5m/\lambda$ and $7m/\lambda$. Note also the important observation that the highest power of frequency in the denominator of the transfer function, which governs the slope on a Bode plot, increases in proportion to the square of the number of degrees of freedom. As a consequence, although the number of resonant peaks and low frequency amplification factors will increase, multi-stage isolators can offer very rapid attenuation of transmitted force and displacement beyond the cross-over frequency.

10.4 Measurement of the viscous damping coefficient

It is very difficult to design an instrument with a closely specified damping factor, so it may need to be measured after building. Also control of existing mechanisms may require knowledge of their damping. It is therefore worth spending some time considering such measurements.

If the damping factor is low (say $\xi < 0.05$) the maximum dynamic amplification factor (or quality factor, Q) can be measured and Equation 10.29 applied directly. Under these conditions the maximum occurs very close to the undamped natural frequency. An alternative method for lightly damped systems is to use experimental signal processing techniques to obtain a plot of the transfer function. The energy dissipated by a system is proportional to the square of its displacement amplitude. Also the points at which the magnitude of $H(i\omega)$ falls to $0.707Q$ are known as the half power points and were of particular interest to early radio pioneers. The frequency bandwidth between these points is related to the damping coefficient by

$$\Delta\omega = 2\xi\omega_n = \frac{\omega_n}{Q} \qquad (10.47)$$

This equation provides a quick method for estimating damping coefficients where there is little dissipation. As the damping increases it becomes necessary to use more analytical methods. The most popular of them is the logarithmic decrement or hammer test, which measures the decay envelope of the impulse response. Equation 10.20 shows the transient response of a second order system to be a sinusiod of period ω_d exponentially decaying with a time-constant $-\xi\omega_n t$. The sinusiodal term has a period t_p, although the response is not strictly periodic, so the ratio of amplitudes after n complete cycles will be

$$\frac{x(t + nt_p)}{x(t)} = \frac{e^{-\xi\omega_n(t + nt_p)}\cos(\omega_d(t + nt_p))}{e^{-\xi\omega_n t}\cos(\omega_d t)} = e^{-\xi\omega_n nt_p} \tag{10.48}$$

Taking the natural logarithm of Equation 10.48, noting that the undamped natural frequency, ω_n, is related to the damped natural frequency, ω_d, by Equation 10.16 and expressing t_p in terms of ω_d, we define the logarithmic decrement, δ, as

$$\delta = \ln\left(\frac{x(t + nt_p)}{x(t)}\right) = \frac{-2\pi\xi n}{\sqrt{1 - \xi^2}} \tag{10.49}$$

Squaring and solving for the damping factor ξ yields

$$\xi = \frac{\delta}{\sqrt{(2n\pi)^2 + \delta^2}} \tag{10.50}$$

This relationship provides a very convenient basis for measurement. A system is excited by an impulse (applied using a hammer or simply by flicking it with a finger) and the decaying response measured on an oscilloscope. The relative peak amplitudes of n successive cycles can then be measured and inserted into Equation 10.50 to determine the damping factor. This method has the advantage that it is a purely ratiometric measurement and, as such, the magnitude of the *input* does not have to be measured. Some skill is need to use this test effectively, but, once mastered, it becomes a standard technique for assessing the amount of damping within a structure. It is important that the initial impulse is not too sharp. If higher harmonics are accidentally excited a rather confusing combination of transients is superposed on the decay curve. This may be observed, for example, if one taps an active component of an instrument with a pencil and with a steel rod. In the latter case, the impulse is shorter (since the material is harder), higher harmonics are excited and the decay envelope distorted from that of the slower pulse. Unfortunately, there is no generalized, quantifiable recipe for hammer testing and one must apply both experience and trial and error to obtain a usable result. As a very general rule, the softer the hammer the better.

10.5 The design of resonant systems: Followers and isolators

It is often relatively easy to decide in the early stages of a design upon the natural frequencies that a system may contain. Suitable values for stiffness and mass can then be explored. The same does not apply to the choice of a suitable damping coefficient for a particular application. In fact, very little has been written on this important topic. One reason for this is that few designs are constructed with an ability to vary the damping coefficient. Another reason is that the specification is somewhat subjective. For example, motor vehicle suspension dampers vary considerably between sports cars and family saloon cars and between countries because of the variable road and weather conditions. Often, therefore, systems are constructed, their performance tested and, if necessary, energy dissipating devices added or modified to rectify undesirable dynamic characteristics.

With smaller systems under closed loop control, it may be simple to include adjustable damping through the use of a derivative element in the feedback loop. This is part of a standard control strategy that combines proportional, integral and derivative feedback to optimize responses. Such design is beyond the scope of this chapter but is concisely covered in the texts of Richards, 1979, and Dorf, 1980. We restrict ourselves to a brief exploration of parameter selection, and notably damping, for second order mechanical systems used for two conflicting purposes

1. to follow dynamically a displacement as accurately as possible:- tracking.

2. to reduce the influence of external and internal disturbances:- vibration isolation.

10.5.1 Stylus tracking

Consider a stylus profiling over a surface. A typical design aim would be to scan as rapidly as possible without the sensor dynamics unduly distorting the fidelity with which the surface is reported. Normally very slow variations must also be measured, so the system is low pass and only frequencies below the first significant resonance can be measured. Thus we tend to design for a high natural frequency. Other design restrictions may limit the extent to which this may be done and we might then consider how to set the damping to provide the best use of the available bandwidth. This is best illustrated by example, such as the atomic force microscope probe shown schematically in Figure 10.9.

For present purposes we assume that a probe of mass m is supported by a 'massless' cantilever beam of transverse stiffness, λ. A high resolution sensor monitors the bending of the beam relative to a reference base. This whole assembly is attached to a stiff servo-drive considered to have a much higher resonant frequency than the beam. To obtain a profile of a specimen surface, the probe is moved into close proximity until van der Waals or contact forces cause the beam to bend (either towards or away from the surface depending on the closeness of the probe tip to the surface). Effects of the near proximity of the

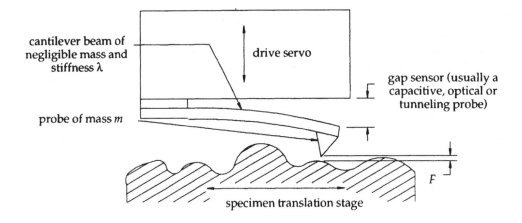

cantilever beam of
negligible mass and
stiffness λ

drive servo

gap sensor (usually a
capacitive, optical or
tunneling probe)

probe of mass *m*

F

specimen translation stage

Figure 10.9 Schematic representation of an atomic force proximity probe.

probe to the surface are likely to influence the stiffness of the system. A discussion of this is beyond the scope of this book. The servo maintains constant sensor output, corresponding to constant beam deflection and so constant separation, while the specimen is traversed. The probe follows the surface and the drive displacement can be considered an accurate representation of the surface profile. However, the very nature of feedback requires a small 'error' signal and so a slight variation in the bending of this beam. Thus the beam-mass system sees an excitation signal of frequency dependent on the surface structure and on the speed at which it is scanned.

The sensors in more conventional stylus measuring instruments are also in effect second order systems. Additionally they rely on the interface forces to displace the stylus. There is a larger direct excitation effect and also the possibility of surface damage induced by high stylus forces. Again, the first resonance would usually be set a high as possible given other constraints such as the unavoidable mass of the sensor element and stiffness being low enough that the contact force does not vary much stylus position. With a contact stylus simple oscillation is rarely a problem but the tendency to jump from the surface (and lose track) at peaks is common and is also related to dynamic amplification. Increasing the damping to reduce the dynamic amplification will lead to increased contact force when the stylus is moving, so there are conflicting requirements. On the one hand, a low damping coefficient is required to limit possible surface damage, on the other, a high damping factor will reduce both lift-off at surface peaks and induced motion of the probe near its natural frequency. The usual compromise is to keep the damping small and to measure slowly.

An alternative compromise would be to look for the value of damping that results in an output adequately representing the measurand (the 'true' surface) over the widest bandwidth. This has been investigated theoretically by Whitehouse, 1988 and 1990, who compared the correlation functions of random surfaces and output signals with models of contact force. It examined how the signal fidelity reduced as frequency components close to the resonance, equivalent to faster traverses over a specimen, were introduced. The closest correlation between surface profile and probe displacement, in both vertical and lateral feature size, occurred with a critical damping factor of approximately 0.6, a little smaller than the value derived from Equation 10.21. This value was derived for both follower-type systems, such as STM and AFM, and for stylus instruments using surface reaction forces. Experimental investigation by Liu *et al.*, 1993, verify that values for the damping ratio of between 0.5 and 0.7 correspond to optimum measurement fidelity for a conventional stylus instrument

10.5.2 Vibration isolation

Seismic isolation is one of the most important aspects of protecting systems from vibration. The basic idea was explored in Section 10.2.4 and Figure 10.5, where it is seen that a second order system responds little to floor vibrations at frequencies significantly above its resonance. A massive, spring mounted table will be more stable than the floor upon which it stands. Generally, one expects that only systems themselves of at least second order will be vulnerable to moderate levels of external excitement. Thus the prevalent situation is that illustrated by Figure 10.6 and analyzed in Section 10.3.1, where the instrument dynamical characteristics are in series with those of the table. The overall response of such a system, Equations 10.43 and 10.44, is illustrated by the examples of Figure 10.7 and Figure 10.8. In all cases the design parameters are the natural frequencies, which depend on the masses and stiffnesses, and the damping factor.

Vibration transmission is a low-pass process, so for the design of a general purpose low vibration table the first priority is to set its natural frequency as low as is compatible with other design constraints. This may be done by increasing the table mass and reducing the stiffness of its supports. Increased damping also reduces the resonant frequency, but not by enough to be useful at reasonable levels. High damping reduces the attenuation at higher frequencies, Figure 10.5, so the critical damping factor should be kept well below unity unless the resonant peak is unavoidably set close to the design stop-band. There are often practical limits to the mass of the table and reducing the support stiffness may cause excessive initial deflection under the dead weight of the table. The lower limit to the stiffness is usually imposed by space restrictions or by the level of deflection with which a human operator feels 'comfortable'. Ideally we require supports that deflect little to initial load and also have low stiffness to incremental load, something not achievable with linear elastic systems. However the extension of a helical spring can be reduced by turning it inside out. This causes residual stresses in the spring that must be overcome by the initial load before it will extend at all, without increasing the spring rate. This allows the resonant frequency to be reduced by up to a factor of two for the same apparent stretch, or, conversely, the same frequency with spring of one quarter the physical length. It is often known as a 'negative length' spring, Pippard, 1985, Nelson, 1991. There is, however, often sufficient space in a laboratory to accommodate large deflections if the springs are anchored to the ceiling and the instrument hung. Using three flexible elastic supports around the periphery of the instrument a trifilar pendulum results. This configuration is semi-kinematically correct and therefore has the maximum number of links (abeit elastic) that will ensure all are in tension. Air bellows supports provide another method of achieving a low frequency table without excessively large displacements. The table is supported on pistons that move within cylinders under the influence of forces caused by vibration of the foundation. Motion of the piston forces air to pass to and fro through a small orifice to act as a damping element. Its advantage is that the natural frequency of the table depends not on the displacement but *only* upon the volume of the air on one side of the piston, the ratio of specific heats for the air and the geometry, Jones (1967). Commercial air-spring systems are readily available. A less efficient, low cost method of vibration isolation, which works on similar principles, is to rest the table on moderately inflated tyre inner tubes.

When anti-vibration mounts for a specific instrument are to be designed, the above guidelines are still valid but there may be a little more scope for design compromise. The instrument dynamics probably restrict the range of frequencies to which it is highly sensitive and satisfactory performance may be obtained providing the mounts do not transmit much in that region. This may be regarded as a case of deliberate mismatching: if the peak responses of two systems are far apart there is little chance of energy being transmitted through one to the other. Designing instruments to have a high natural frequencies is a major contribution to the provision of good vibration immunity. For layouts such as Figure 10.6 the resonant frequency of the support (including the mass of the instrument) must be less than one half, and preferably an order of magnitude or more smaller if the damping is low, of the lowest frequency that is to be isolated. That, in turn, will be rather less than the lowest natural frequency of the instrument, compare Figure 10.7 and Figure 10.8. Increasing the mass of the table and reducing that of the instrument while maintaining the original stiffnesses (or, even better, also reducing the stiffness of the support and increasing it in the instrument) reduces the residual displacement at the instrument sensor. Additionally, as the resonances move further apart, the two systems have little influence on each other and for analysis purposes can be treated as isolated without excessive error. This simplifies design calculations since the two natural frequencies are then found from the two independent stiffness to mass ratios.

The best choice of damping factor depends upon the separation of the natural frequencies of the table and instrument. It is best kept low to provide rapid attenuation at higher frequencies, provided that it is sufficient to prevent a dynamic amplification near the table resonance so large that significant energy is coupled across the mismatch. As the damping is increased this error is reduced at the expense of an increased sensitivity to frequencies away from the resonance. If the mass of the instrument is much less than that of the table, a good compromise is a critical damping factor in the region of 0.1, or somewhat less, Snowden, 1968. This will kill low frequency vibrations relatively quickly without significantly reducing the attenuation at higher frequencies. As a general rule, light damping and a difference in natural frequencies of at least three orders of magnitude offers sufficient isolation for most high precision applications. This might be used as an initial, perhaps rather severe, design target.

Seismic isolation is only part of vibration protection for there are many possible routes whereby vibrations can bypass the isolator. Often an instrument will be connected to the outside world by cables and pipes. These can transmit over a wide frequency bandwidth. They should be made as flexible as possible and clamped rigidly to the table at two separate points. Two point fixing is necessary because a single clamp has little torsional rigidity in axes perpendicular to the cable and so leaves it free to vibrate laterally. The clamped section can be treated as a fixed end beam for the estimation of its natural frequency of lateral vibration and so may be designed to mismatch other major resonances. Another source of disturbance can arise from elastic waves travelling along the pipe material. These are often of high frequency and so may be very damaging to instrument fidelity. This can also occur within the table suspension and is a well known problem associated with metallic spring supports, Snowden, 1968. The only way of attenuating these vibrations is to insert a material having a high internal loss factor (usually rubber) in the force path, coupled to the cables or spring suspension. Finally, in hydraulic systems noise from the pumps and control valves may be transmitted through the fluid. The commonest attenuation technique is to use an accumulator, a system analogous to a bellows support. It consists of a large tank containing an air filled bag. Small ripples in the hydraulic pressure are attenuated by compression and expansion of the bag. Although effective for small, high frequency vibrations, lower frequencies and high amplitude surges will still be transmitted to the instrumentation.

Isolation over almost any range can, in principle, be achieved by active feedback, although it is often a very complex and expensive technique. For the models presented here, active isolation is simple to implement. It requires only the insertion of a velocity sensor and a force actuator in parallel between the floor and the table. The sensor signal is used to drive the actuator and, as long as the force actuator has opposite sign to the velocity, energy will be extracted from the table. Unfortunately, most instruments do not behave like the single degree of freedom model. Real structures vibrate with a wide range of modes, frequencies and amplitudes. To achieve effective isolation, gauges and actuators must be located near the points of maximum amplitude (antinodes). This considerably complicates the design. Having initially determined the optimal points for mounting the controllers, it is commonly found that attaching them modifies the vibrational mode pattern of the system. With a poor quality control signal, the actuators can inject net energy into the system. Practical active isolation is beyond the scope of this book.

10.5.3 Noise isolation

Finally, precision systems may need protection from the influence of airborne disturbances, often called noise to distinguish them from vibrations transmitted through a structure. The strategies are either to convert sound into heat through acoustic absorption or to reflect it away from the instrument. Fortunately, sound waves usually have little energy and can then be absorbed by cheap casings made from polystyrene or fibrous materials, or cloth surrounds, although these may be inconvenient. For casings not required to support appreciable loads, compressed fibre panels are cheap and absorb sound well since multiple internal reflections lead to high frictional dissipation. However, for very high attenuation it is necessary to first reduce the energy that must be absorbed. Thus typical noise isolation casings are sandwich structures, with an outer skin that reflects much of the energy before it reaches the absorber.

The sandwich structure functions primarily by providing series acoustic impedances that are highly mismatched so as to promote reflection, rather than transmission, of energy at the interfaces. Air has low density, so the outer skin should be of high density and the inner absorber again of low. Energy transfer is aided by scattering and multiple interactions with rough surfaces so the outer skin should be smooth, perhaps polished or coated with a hard enamel or two-part epoxy paint. Some energy will always be transferred to the outer layer and to minimize its coupling to the absorber, we wish its resulting vibration to have as low an amplitude as possible. If we model the outer layer as a linearly elastic material vibrating sinusoidally as a plate, the input energy could couple to either the potential or kinetic energy storage components, which are a

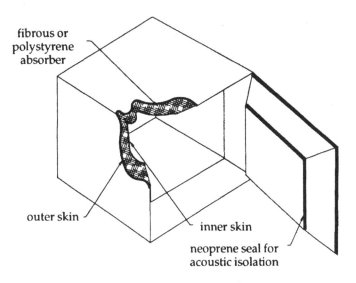

fibrous or
polystyrene
absorber

outer skin

inner skin

neoprene seal for
acoustic isolation

Figure 10.10 A basic laboratory acoustic isolation chamber

quarter of a cycle out of phase. Thus, from Equation 10.1, the maximum amplitude of the plate vibration for a given input energy is related to the square roots of its mass and its stiffness. Consequently, the ideal material for the outer skin will have both high density and high elastic modulus, to provide so called 'mass damping'. Tungsten alloy is often used for very demanding applications. More typically, mild steel will be perfectly adequate and even painted wood or acrylic is useful. Normally the thick absorber layer is covered by an inner skin that makes construction easier and also protects against the generation of dust particles. The outer material should not be used on the inside because noise should not reflect internally. The inner skin should be a low density, relatively absorbent material, such as thick cloth, rubber or wood. A typical low-cost laboratory instrument housing is shown in Figure 10.10. Note that sound waves will diffract through holes to fill the enclosed volume, so the sound proofing should, if possible, be air-tight. Door seals should be used and essential wires, pipes, drive shafts, etc. should be brought into the cabinet through 'silently sealed' ports.

10.6 Damping of small vibrations

Having discussed the specification of damping factor in the previous sections, we here look briefly at the types of damper available to the instrument designer. A more comprehensive overview of the conventional types of damper and damping material is provided by the excellent handbook of Harris and Crede, 1961. It should be stressed that there is considerable controversy over the extent to which conventional methods apply to high precision applications. There is little or no evidence to support, or deny, the continued efficiency at the nanometre level of the damping mechanisms discussed in this section. Failure to do so could explain the often reported need to resort to active techniques. Such doubts do not, however, preclude their use in initial vibration attenuation mechanisms.

The simplest and cheapest damper is just a rubber spring. Its very complex molecular structure causes its stress-strain curve to exhibit considerable hysteresis, which increases with a reduction in cross-linking. The area enclosed within the hysteresis loop measures the amount of energy dissipated during each cycle. This known to increase with both amplitude and frequency, so they may be relatively ineffective for very high precision instrumentation mechanisms. Indeed, it has proved very difficult to control instrument mechanisms at the nanometre level unless their dominant resonant frequency is greater than 1kHz. Rubbers (elastomers) or similar viscoelastic materials generally lose their damping efficiency as temperature is reduced. For the majority of rubbers this begins at approximately 70 °C and they cease to be useful for damping at below –10 °C. At low frequency, the molecules can curl and uncurl at roughly the same rate at which the stress is changing. There is then very little phase difference between stress and strain and the damping tends to decrease. Typically the efficiency of damping in rubber springs drops rapidly below approximately 100 Hz. Below this frequency the critical damping factor might be as low as 0.01 for some natural rubbers, in the region of 0.1 for most elastomers but up to 0.5 or more for some very low cross-linked butyl rubbers (polyisobutylene). The materials with high internal damping are very sensitive to temperature variations, Lazan, 1968, Snowden, 1968.

In most solid materials, internal energy dissipation is caused by thermoelastic relaxation, friction at the relative motion between grain boundaries and the migration of dislocations and interstitials. Nearly always the damping energy is very small compared to that of elastomers and the effects are not exploited in isolators. However, material interface damping is used in several forms. If two components are clamped together and then subjected to a stress across the contacting areas, there will be a finite amount of slip. This gives rise to frictional forces that will dissipate energy in the form of heat, particularly if the main load can be carried in shear. The rubbing surfaces may generate and trap oxide particles which further abrade the interface and result in the rapid wear commonly known as fretting. Good materials selection and low amplitude vibration will minimize the problems of fretting, but the application to damping may still be limited. Hence it is common to separate the surfaces by a thin layer of viscoelastic rubber or epoxy adhesive to provide interface damping. Materials having a high density of large internal interfaces will also tend to absorb energy. For example, high carbon content cast iron can have a damping factor as high as 0.01 and a manganese, copper, aluminium alloy has one of about 0.012, Langham, 1967. Similar internal interfaces can be created in composites, for example by mixing stones of the correct size distribution into a resin to form a polymer-concrete such as 'Granitan', Renker, 1985, see also Chapter 8. Although it has frequently been used for the base of precision machines, there is little trustworthy data about its damping properties, see, for example, Weck and Hartel, 1985. The idea of designing a composite material for its damping properties is relatively new. However, interfacial damping has occurred as a bonus, for example in probe systems for surface profiling instruments and coordinate measuring machines made from rods of carbon fibre reinforced plastics primarily because of the low negative thermal expansion coefficient of the fibres (approximately -0.2 to -0.4×10^{-6} K^{-1}). Carbon or glass fibre reinforced plastics are also good materials for instrument cases because of their acoustic absorption and thermal insulating properties. The last type of interface damping to be considered is that provided by a volume of fine powders. If a hollow section is filled with particulates, they will slide over each other upon the application of small forces. Typically an extruded metal optical bench would be filled with sand to 'dull' its response, and, incidentally, to increase its mass. The dissipation is again provided by Coulomb frictional forces. The forces required to induce interface shear are much smaller that in solids because of the low inertial mass of the individual particles. It is not clear whether there is a minimum amplitude for their effectiveness.

Stand-alone dampers, that is dissipating mechanisms, are effective for many conventional applications, but may require more relative motion than can be tolerated in our context. Magnetic damping, for example through eddy-currents, tends to fall in this category. Magnetoelastic materials (see also Section 7.1.6) provide a single material equivalent that may be better at small amplitudes. Strain results in their magnetic domains being irreversibly distorted, causing a net energy loss that provides mechanical damping. Nivco (72% Ni, 23% Cr) is of particular interest. It has a loss coefficient that increases linearly with applied stress, is significantly higher than almost all metals at low amplitude and, for the majority of engineering applications, can be considered independent of frequency, Harris and Crede, 1961.

A more common type of damper uses the relative motion between two mechanism elements to pump a fluid through an orifice. Viscous drag produces a damping characteristic that can be adequately modelled by a classical linear force-velocity relationship. Both gases and liquids are used in this type of device. Since gasses are compressible, they may be used as both damper and support spring, see Section 10.5.2. Oil, or other liquid, dampers often offer the cheapest method of energy dissipation but they must be used in conjunction with springs. In principle these devices will operate at very small amplitude, but the pistons used to pump the fluid through the orifice can couple vibrations by elastic wave propagation and there will be limitations due to the mean free path of molecules in the fluid.

For low amplitude vibrations in mechanisms having small, low mass components (say, to a few grams and millimetres) the surface tension of a liquid can provide damping without any coupling other than a meniscus across the gap separating two components. Typically a spot of silicone oil would be used, not only for its viscosity but because its very low vapour pressure prevents untoward evaporation. In very small systems there need be no coupling other than by the air in the gap separating two components. Damping, known as squeeze film action, occurs as energy is taken from the system in pumping the fluid in and out of the gap. The degree of damping depends upon the fractional volume change for a given displacement δ. For two parallel plates of constant overlap area separated by a gap δ_o, the volume moved, V, is given by

$$\frac{V}{V_0} = \frac{\delta}{\delta_0} \tag{10.49}$$

Thus the damping force is proportional to both area and the reciprocal of the initial separation, with a sensitivity increasing with the inverse cube of the separation, Gross, 1962. Thus for maximum damping the gap between components should be as small as possible. (There is some folklore that this does not apply to micrometre gaps where damping reduces again as air is excluded from the gap. This probably derives from a misplaced analogy with dielectric behaviour in small gaps. No experimental evidence is openly available.) There are two possible damper configurations; the shear type in which the fluid or air is sheared by the parallel motion of the plates and the squeeze film type exploiting perpendicular motion. Shear damping has the advantage that the damping force is linear with velocity for low Reynold's number flows and amplitude of relative motion is not limited by the gap size. Squeeze film types more readily provide high critical damping factors.

10.7 Case studies

Few are the cases where only dynamical considerations are important in instrument design. Thus we end this chapter, and the book, with brief discussions of two extremely sensitive instruments in which both the design of second order systems and vibration isolation are critical issues. We do not, however, restrict ourselves to those aspects.

10.7.1 An atomic force probe

The importance of dynamical response to the fidelity of atomic force, and similar, scanning microscopies was briefly explored in Section 10.5.1. We consider here the implications of a particular design of transducer for a fine force profiler, Figure 10.11, by Howard and Smith, 1992. To recap, the essence of such devices is the measurement of extremely low forces, attractive or repulsive, between the probe tip and a proximate surface. If the surface is then scanned under the probe, variations in the specimen profile will cause corresponding changes in the separation and thus in the force gauge output. If this signal is fed back to a piezoelectric actuator supporting the probe through a proportional plus integral controller, the probe will effectively skate over the surface maintaining a constant force and constant separation. Monitoring the extension of the piezoelectric actuator, for example with a capacitance gauge, can give a measure of the profile at a constant force of a few tens of nanonewtons. Figure 10.11 shows the assembly that mounts to the piezo-actuator. The base is a Zerodur optical flat having a small etched step (approximately 8 μm) on which a metallized film forms one electrode of a capacitance gauge. A thin silica cantilever beam, approximately 6 mm long 2 mm wide and 100 μm deep, is bonded onto the base, above and overhanging the step. This has a metallized rear face which forms the second electrode of the gauge. A diamond probe tip protrudes from near the free end of the cantilever.

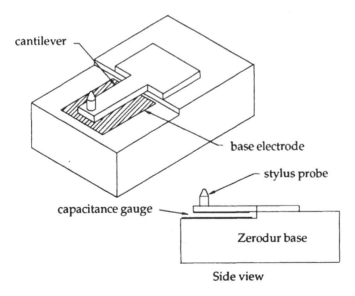

Figure 10.11 Basic force probe.

As the actuator lowers the probe toward a specimen, the interfacial forces between the probe and surface cause a bending of the beam which is monitored by the capacitance gauge. The structure behaves closely as a second order system with a natural frequency of around 800 Hz.

This system is rather more bulky overall than some other designs, but has a number of advantages.

1. The resonant frequency is sufficiently high and the base massive enough to make resonance mismatching for vibration isolation relative simple. Note that this is only possible because the extreme displacement sensitivity of the guaging provides nanonewton sensitivities in relatively stiff beams.

2. The high aspect ratio in the electrode gap of the cantilever gauge provides significant air damping. Its value can be predicted theoretically over a wide range of damping factors, and present designs are near to critically damped. Such designs are remarkably insensitive to external vibration sources.

3. The fabricated design allows metallization on both faces of the cantilever to compensate for bimorph effects due to differential thermal expansion between the cantilever and electrode. This reduces thermally induced variations of the interface force within the measurement loop. The cantilevers may also be preannealed to reduce internal stresses. (Most probes are made using standard lithographic processing which often produces cantilevers having unknown internal stresses and serious bimorph instabilities).

4. The gauge can be built as a teeter-totter device, enabling differential gauging with the reference capacitor operating in anti-phase with the working gauge, see Figure 10.12.

Early scanning tunnelling microscopy was cursed by vibration sensitivity and its history shows several stages of development in both instruments and vibration isolation. Original isolation methods included levitating instruments on superconducting coils or hanging them from a trifilar pendulum on long elastic wires with viton rubber isolators at each end to attenuate high frequency acoustic waves, Binnig et al., 1986. Viton rubber was favoured mainly because it has a vapor pressure of approximately 10^{-8} and is therefore suitable for moderately high vacuum applications. It was soon realized that simple multi-stage vibration isolation systems would be sufficient and much less complex. A very effective version simply stacked successive stainless steel plates separated by Viton rubber spacers, Gerber et al., 1986. Interestingly, an order of magnitude increase in stability was observed with the complete system enclosed in a vacuum chamber. It is now widely recognized that such high frequency instrumentation is susceptible to electronic and acoustic noise. Modern commercial STMs are much less susceptible to vibration than the early experimental ones, mainly because good design of small metrology loops makes them less efficient as 'vibration antennas'.

10.7.2 A gravity wave antenna

Some of the greatest challenges for super precision instrument engineering are to be found in gravity wave detectors. Magnetized neutron stars known as pulsars emit a sharp pulse of radio emission upon each rotation. Some binary systems are observed to slowly approach one another and the slow-down rate of their pulsing can be accurately modelled by the emission of gravitational waves. If they could be measured, the general theory of relativity could be tested with a high degree of accuracy and it may enable astronomers to monitor the creaks and groans of the expanding universe (for an elementary introduction see Wheeler, 1990). The waves are totally void of mass-energy and constitute a ripple in space time that is capable of transferring energy to a mass. One proposed method of detection is to suspend a mass and observe its distortion as the wave passes through. The strain in the rod then excites the resonant frequency of this 'antenna' which, for a number of reasons, is chosen to be around 1 kHz. Magnitudes of the expected strain are *minute*. Present capacitance sensors are capable of detecting strains of the order 10^{-18} and the latest goal is 10^{-20}, Linthorne *et al.*, 1990. This corresponds to one attometre change in the length of a 100 m bar. Clearly, science is not be content with mere

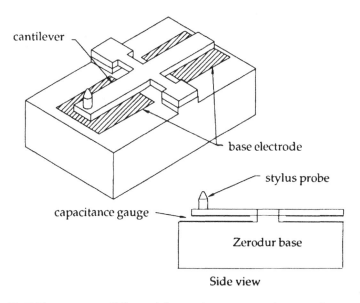

Figure 10.12 Teeter-totter differential capacitance gauge force probe.

nanotechnology! Typical gravity antennae consist of aluminium cylinders of perhaps 20 tonne mass. An extreme degree of vibration isolation in the measurement bandwidth is essential to the success of projects taking place in America, Japan and Western Australia. Superpositions of series connected spring-mass-damper sections are used for the antenna support. Exploiting the fact that for each individual isolator the attenuation follows an inverse square law above its resonant frequency, a 300 db attenuation has been achieved with a section natural frequency of 100 Hz. It has been pointed out, Debra, 1989, that a natural frequency of 1 mHz would be required if a single stage were used. This would be impracticably floppy.

Another important aspect of the isolator design is that, to avoid confusion between gravity waves and a mechanical impulses in the support structure, individual elements in this passive system must self-excite only at frequencies well above the 2 kHz bandwidth requirement. Much care is taken to select springs and masses for the isolation support that do not contain natural resonances below this value.

It is one of the great triumphs of precision electronics and precision mechanics working together, that we even contemplate practical gravity wave detectors.

References

Binnig G., Rohrer H., Gerber Ch. and Weibel E., 1982, Tunneling through a controllable vacuum gap, *Appl. Phys. Letts.*, **40**(2), 178-180.

Debra D., 1989, *Vibration isolation*, presented at the 5th Int. Precision Engineering Seminar, Monterey, California

Dorf R.C., 1980, *Modern Control Systems*, Addison Wesley, London. (For a discussion of the choice of damping coefficients see pages 110-119).

Gerber Ch., Binnig G., Fuchs H., Marti O. and Rohrer H., 1986, Scanning tunneling microscope combined with a scanning electron microscope, *Rev. Sci. Instrum.*, **57**(2), 221-224.

Gross W.A., 1962, *Gas Film Lubrication*, John Wiley & Sons,Inc., New York, Chapter 6.

Harris C.M. and Crede C.E., 1961, *Shock and Vibration Handbook, Volume II*, Chapters 30-37, McGraw-Hill, London.

Howard L.P. and Smith S.T., 1992, Long range constant force profiling for measurement of engineering surfaces, *Rev. Sci Instrum.*, **63**(10), 4289-4295.

Jones R.V., 1967, The measurement and control of small displacements, *Bulletin of IoP*, 325-336.

Landau L.D. and Lifshitz E.M., 1960, *Mechanics*, Vol. 1 of *Course of Theoretical Physics*, Pergamon Press, Oxford

Langham J.M., 1967, A new high damping alloy, *Proc. 17ᵗʰ Annual Pacific Conference of the American Foundrymens' Society.*

Lazan B.J., 1968, *Damping of Materials and Members in Structural Mechanics*, Pergamon Press, London, Chapter 3.

Linthorne N.P., Veitch P.J. and Blair D.G., 1990, Interaction of a parametric transducer with a resonant bar gravitational radiation detector, *J. Phys. D.*, **23**, 1-6.

Liu X., Chetwynd D.G., Smith S.T., and Wang W., 1993, Improvement of the fidelity of surface measurement by active damping control, *Meas. Sci. Technol.*, **4**, 1330-1340.

Nelson P.G., 1991, An active vibration isolation system for inertial reference and precision measurement, *Rev. Sci. Instrum.*, **62**(9), 2069-2075.

Newland D.E., 1990, *Vibration Analysis and Computation*, Longman, London.

Pippard A.B., 1985, *Response and Stability: An Introduction to the Physical Theory*, Cambridge University Press, Cambridge.

Renker H.J., 1985, Stone-based structural materials, *Precision Engineering*, **7**, 161-164.

Richards R.J., 1979, *An Introduction To Dynamics and Control*, Longman, London.

Snowden J. C., 1968, *Vibration and Shock in Damped Mechanical Systems*, J. Wiley and Sons, London, Chapter 1.

Weck M. and Hartel R., 1985, Design, manufacture and testing of precision machines with essential polymer concrete components, *Precision Engineering*, **7**, 165-170.

Wheeler J.A., 1990, *A Journey Into Gravity and Spacetime*, Scientific American Library, New York.

Whitehouse D.J., 1988, Revised philosophy of surface measuring systems, *Proc. Instn. Mech. Engrs*, **C202**, 169-1.169-185.

Whitehouse D.J., 1990, Dynamics aspects of scanning surface instruments and microscopes, *Nanotechnology*, **1**, 93-102.

APPENDIX A

Airy point mounting of datum bars

Ideally, beam structures should be supported such that distortions are minimized. Obviously, care must be taken to avoid unacceptable distortions due to locally applied loads. If they are static, appropriate positioning of the supports can be calculated, but difficulties always remain with moving loads. Distortion due to self weight is also of importance, for example in datum bars (or bookshelves). One special case is the mounting of metrological straightedges or length bars, which may be considered as uniform horizontal beams held on two simple supports. The problem of supporting long prismatic rods of length L was first addressed, for length standards, by George Airy. It is now established practice that when parallelism of the end faces is required, the two supports should be placed symmetrically at a separation of $0.577L$. Correspondingly, for minimum deviation from straightness, the supports should spaced at $0.554L$ giving equal sagitta at each end and at the centre, Figure A1. These are known as the Airy points, sometimes informally called the 'two ninths points'. These support configurations will give optimum performance in datum applications, permitting a lighter construction than otherwise possible.

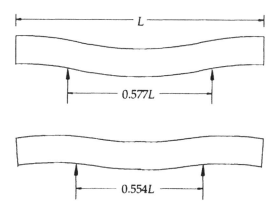

Figure A1   Supporting length bars and straightedges

AUTHOR INDEX

SUBJECT INDEX

A

Abbe alignment principle, 71, 91, 257
Abbe errors
  See Errors
  in slideways, 262
Abbe offset, 70-71, 224
Accuracy
  definition of, 6
  of actuators, 182
Acoustic noise, 312
  impedance mismatching, 312
  low cost isolation chamber, 313
Actuation force
  line of, 98
Actuators
  nonlinearity, 78
  properties of, 182
  summary of performance, 201
  types of, 181
Adhesives
  applications in high stability
    structures, 241
Age hardening, 235
Air bearings
  See Hydrostatic bearings
Airy points, 223, 323
Alignment, 44
  See errors
  principles of, 70
Alumina, 239
  property profile, 239
Aluminium, 224

property profile, 234
Amonton's laws, 60
Angle spring hinges, 117
  crossed strip and monolithic
    hinge, 119
  cruciform, 120
  stability of axis, 119
Ångstrom ruler
  See x-ray interferometry
Angular gearboxes
  See rotary speed converters
Angular levers, 162
  insensitivity to environment, 162
  See also Pulley drive
  See also Wedge
Areas under the normal curve, 12
Assay balance, 73, 84
  limits of, 148
Assemblers, 88
Atomic force microscope (AFM), 75, 211
Atomic force microscopy
  See Microscopy
Atomic force probe
  case study, 317
Averaging
  compliant nut or Merton nut, 78
  in elastic design, 60
  mechanical, 188
  through overconstraint in
    flexures, 117

M

T

U

V

W

T - #0625 - 101024 - C0 - 246/189/20 - PB - 9782884490016 - Gloss Lamination